Horizons
Mathematics

Ellen Carley Frechette

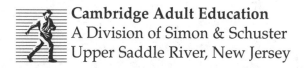

Cambridge Adult Education
A Division of Simon & Schuster
Upper Saddle River, New Jersey

Executive Director: Mark Moscowitz
Project Editors: Robert McIlwaine, Bernice Golden, Laura Baselice, Keisha Carter, Lynn Kloss
Writer: Ellen Carley Frechette
Series Editor: Roberta Mantus
Consultants/Reviewers: Marjorie Jacobs, Cecily Kramer Bodnar
Production Manager: Penny Gibson
Production Editor: Nicole Cypher
Marketing Manager: Will Jarred
Interior Electronic Design: Flanagan's Publishing Services, Inc.
Illustrator: Accurate Art, Inc. & Andre V. Malok
Electronic Page Production: Flanagan's Publishing Services, Inc.
Cover Design: Armando Baez

Printed in the United States of America
1 2 3 4 5 6 7 8 9 10 99 98 97 96 95
ISBN: 0-8359-4633-9

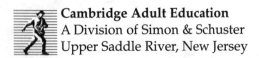

Cambridge Adult Education
A Division of Simon & Schuster
Upper Saddle River, New Jersey

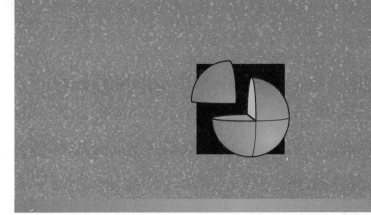

Contents

Unit 3 **Fractions** **70**

Unit

1

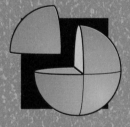

Whole Numbers

Place Value

What You Know   Which would you rather have — $123, $231, or $321?

You probably know that 321 is the largest of these numbers. How did you know this? These three numbers use the same digits — 3, 2, and 1. A **digit** is a number from 0 through 9. Even though these three numbers use the same digits, they have different values because the digits are in different places. So, to understand our number system, you need to understand *place value*.

How It Works   We use the **place value** system to read and write numbers. The value of a digit depends on what place it holds in a number.

Let's look at the place value chart:

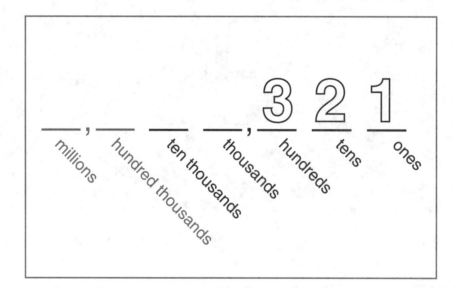

In the number 321, the 3 is in the hundreds place. This means that it has a value of 300 (3 hundreds). In the number 123, the 3 is in the ones place. Therefore, the value of the 3 in this number is 3 ones, or 3.

What is the value of the 3 in 231? It has a value of 3 tens, or 30.

There are larger place values than the ones you see in the chart on page 2. For example, you may be familiar with hundred millions and billions. In this book, you'll be working mostly with numbers that do not go over the ten thousands.

Try It Use the chart on page 2 to help you in your understanding of place value.

What is the value of the 2 in the number 28,319? _____

The 2 is in the **ten thousands** place. Its value is 2 ten thousands, or 20,000.

In the number 4,586, what is the value of the 5? _____

The 5 is in the **hundreds** place. Its value is 5 hundreds, or 500.

In the number 567, what is the value of the 7? _____

The 7 is in the **ones** place. Its value is 7 ones, or 7.

Practice What is the value of each underlined digit below? (The first answer is given.)

1. 7̲85 _700_	4,1̲20 _____	8,908 _____
2. 9̲8 _____	2,0̲34 _____	1̲3,762 _____
3. 3̲0 _____	3̲ _____	3̲03 _____
4. 29̲0 _____	1̲88 _____	5̲6,001 _____
5. 20̲0 _____	1,5̲09 _____	6̲9,815 _____

Write the following numbers using the correct digits and place values. (The first answer is given.)

6. nine hundred thirty-five   _935_

7. seventy-seven   _____

8. eight thousand three hundred five   _____

9. fourteen thousand   _____

10. forty-four thousand six hundred ten   _____

Answers are on page 114.

Follow-Up. On ten separate slips of paper, write the digits 0 through 9.

By yourself or with a partner, put these numbers in the order that makes the *largest* possible number. Then use all the numbers to make the *smallest* possible number. Then write out both numbers in words.

Using a Calculator

What You Know   Have you ever used a calculator to do your work? Compare your calculator to the one pictured below. Your calculator may look slightly different from this one. Although it may look different, it doesn't matter. It's what the calculator can do that's important. Most calculators have the features shown below.

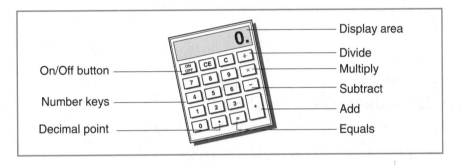

First, let's see what happens when you turn on your calculator.

Now let's see what happens when you press a series of numbers.

How It Works   A calculator will not tell you what number keys you should press. It will also not tell you whether to add, subtract, multiply, or divide. Calculators cannot think or make decisions for you. Only *you* have that kind of thinking power.

But a calculator will add, subtract, multiply, or divide for you. For example, if you press the correct keys, it will very quickly tell you what 13 added to 25 equals. Let's try it to see how it works.

When You Press These Keys	Your Calculator Displays
C (to clear the display)	0.
1	1.
3	13.
+	13.
2	2.
5	25.
=	38.

Try It   Using your calculator, add the following numbers.

20 + 98 =   _____

135 + 16 =   _____

201 + 88 + 10 =   _____

If you answered 118, 151, and 299, you have used your calculator correctly. If you did not get these three answers, try entering the addition problems again.

 Math Tip   It's usually a good idea to do problems twice when you use a calculator. It is often easy to press the wrong key by mistake. Always check your work, especially if you are adding more than two numbers.

Practice   Use your calculator to add the following numbers.

1. 209 + 876 =   _____

2. 1,908 + 25 =   _____

3. 145 + 88 =   _____

4. 1,209 + 219 + 12 =   _____

5. 3,001 + 13 + 140 =   _____

Answers are on page 114.

Rounding and Estimating Whole Numbers

What You Know   Suppose that you have $49 in your pocket. Fill in the answer blank below:

In my pocket, I have about _____ dollars.

If you filled in a number that was close to, but not exactly, $49, you already know something about rounding and estimating whole numbers. You may already know that $49 can be rounded off to $50.

How It Works   When you **round off** a number, you are just finding a whole number that is close to that number and easy to work with. Numbers that end in zero are usually easy to work with.

Using a number line is a good way to see how rounded numbers work. On the number line below, can you see that 49 is between 40 and 50?

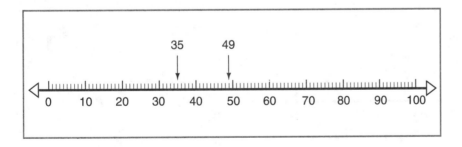

You should also be able to see that 49 is closer to 50 than it is to 40. Therefore, you can say that 49 rounded to the nearest 10 is 50.

 Math Tip   When a number is *exactly* between two numbers, round up to the higher number. For example, 35 should be rounded to 40, not 30.

Rounding to the nearest hundred or thousand works in the same way. You are just deciding what the closest hundred or thousand is to your number.

Example   What is 257 rounded to the nearest hundred?

> ***Step 1***   Between what two rounded hundreds is 257?
> 257 is between 200 and 300, right?

> ***Step 2***   Is 257 closer to 200 or closer to 300? Imagine a number line. 250 is exactly halfway between 200 and 300, so 257 is a little closer to 300 than it is to 200.

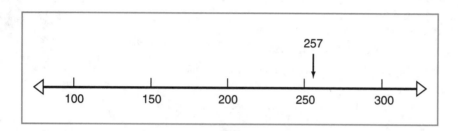

> ***Step 3***   Round the number.
> 257 rounded to the nearest hundred is 300.

Try It   Use a number line to answer the following questions.

The number 23 is closer to 20 than it is to 30. Therefore, 23 rounded to the nearest ten is _____.

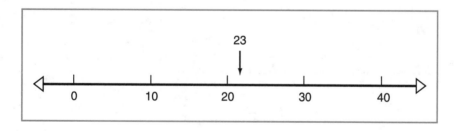

The number 15 is halfway between 10 and 20. What is 15 rounded to the nearest ten? _____

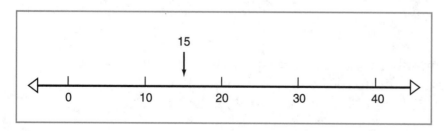

Now try these two by drawing your own number line, or by just picturing a number line in your mind.

Round 385 to the nearest hundred. _____

Round 3,102 to the nearest thousand. _____

If you wrote 20, 20, 400, and 3,000 in the answer blanks on page 8 and above, then you understand rounding!

Practice   Round the following numbers to the nearest ten. (The first answer is given.)

1. 28   __30__   91 _____   89 _____   17 _____

2. 9 _____   52 _____   14 _____   71 _____

Round the following numbers to the nearest hundred.
(The first answer is given.)

3. 201   __200__   98 _____   578 _____   230 _____

4. 125 _____   529 _____   245 _____   79 _____

Round the following numbers to the nearest thousand.
(The first answer is given.)

5. 1,500   __2,000__   1,475 _____   9,008 _____   3,609 _____

6. 1,298 _____   5,550 _____   4,150 _____   8,099 _____

Answers are on page 114.

Follow-Up. Empty the money you have in your pockets right now, or find the balance in your bank account. Now round off this amount to:

1. The nearest dollar

2. The nearest ten dollars

3. The nearest hundred dollars

Adding Whole Numbers

What You Know   You are probably already familiar with **basic addition facts**. In other words, you can add single-digit numbers together fairly quickly to find the **sum**, or total amount. For example, how many tires does a car have, including the spare? You probably thought to yourself, "4 tires plus 1 spare equals 5 tires." This is an addition fact.

To get warmed up, try finding the following addition facts:

3 + 9 = _____   4 + 5 = _____   2 + 8 = _____

Once you are comfortable with adding single digits, you are ready to add larger numbers. Did you get 12, 9, and 10 for the addition problems above?

How It Works

Example 1   Add 28 + 40.

Step 1	Step 2	Step 3
Line up the digits so that ones are under ones, tens under tens, etc. 28 + 40	Add the ones column. 28 + 40 8	Add the tens column. 28 + 40 68

Use the same steps to add more than two numbers.

Example 2   What is 221 + 23 + 101?

Step 1	Step 2	Step 3
Line up the digits so that ones are under ones, tens under tens, etc.	Add the ones column.	Add the tens column and the hundreds column.
221 23 + 101	221 23 + 101 5	221 23 + 101 345

Always read addition problems carefully. Add only the numbers you need to add. Watch out for extra numbers in a problem.

Example 3   Tamika read 36 pages on Saturday and 13 pages on Sunday. She spent 55 minutes reading in all. How many pages did she read altogether?

  36
  13
+ 49 pages   Do *not* add the *minutes* she spent reading.

Many times in this book and in your everyday life, you do not need to find an exact answer to an addition problem. Using round numbers can help you find a quick *estimate* to an addition problem.

 Math Tip   The symbol ≈ means *about* or *approximately* — in other words, close but not exactly.

Example 4   Roberto drove 17 miles yesterday and 51 miles today. *Approximately* how many miles did he drive altogether?

17 rounds to ⟶ 20
51 rounds to ⟶ 50
20 + 50 = 70

Roberto drove 70 miles.

11

Try It Add these numbers. First round and estimate. Then find the exact answer. Use the steps on page 11 if you need help.

35 + 61 _____   35 + 61 = _____

If you round 35 up to 40 and 61 down to 60, your estimate should be 100. The exact answer for 35 + 61 is 96.

101 + 37 _____   101 + 37 = _____

If you round 101 to 100 and 37 to 40, your estimate should be 140. The exact answer to 101 + 37 is 138.

Practice Practice finding these addition facts.

1. 4 + 8 = _____   5 + 6 = _____   3 + 7 = _____   2 + 6 = _____

2. 9 + 7 = _____   4 + 5 = _____   1 + 8 = _____   9 + 5 = _____

Add the following numbers.

3.
$$
\begin{array}{r} 34 \\ +15 \end{array} \qquad
\begin{array}{r} 75 \\ +12 \end{array} \qquad
\begin{array}{r} 90 \\ +\ 9 \end{array} \qquad
\begin{array}{r} 12 \\ +47 \end{array} \qquad
\begin{array}{r} 17 \\ +81 \end{array}
$$

4.
$$
\begin{array}{r} 83 \\ +\ 6 \end{array} \qquad
\begin{array}{r} 61 \\ +18 \end{array} \qquad
\begin{array}{r} 44 \\ +25 \end{array} \qquad
\begin{array}{r} 10 \\ +88 \end{array} \qquad
\begin{array}{r} 11 \\ +56 \end{array}
$$

Estimate an answer to the following addition problems by rounding each number to the nearest ten or hundred.

5. 36 + 99 _____   14 + 28 _____   37 + 11 _____

6. 329 + 790 _____   210 + 66 _____   491 + 550 _____

Use a calculator to add the following.

7. 1,908 + 653 = _____   456 + 2,130 = _____

8. 389 + 3,008 + 65 = _____   2,176 + 3,459 + 1,001 = _____

Solve the following problems. Remember to add only the necessary numbers.

9. Maureen poured 32 fluid ounces of baby formula into 32 fluid ounces of water. How much liquid did she have altogether?

_____

10. Anthony drove 18 miles on Monday and 11 miles on Tuesday. His wife drove the same car 13 miles on Wednesday. How many miles did Anthony drive?

_____

11. Of the four children in his family, Sam is the shortest, at 51 inches. His sister Janice is 8 inches taller than Sam. How many inches tall is Janice?

_____

12. Vera paid $27 for a pair of jeans, $40 for a sweater, and $12 for a blouse. How much did she spend altogether?

_____

Answers are on pages 114 to 115.

Follow-Up. Find a clothing store advertisement in your newspaper. Write a "dream shopping list," including prices, rounded to the nearest $10. Trade your list with a partner and ask your partner to add up the cost. Who spent more money?

Adding and Carrying

What You Know   Sometimes the sum of the numbers in a column adds up to more than 9. In other words, you may get a 2-digit number when you add the ones or tens column. What do you do then?

When working with money, you know that you can exchange ten single dollars (10 ones) for a ten-dollar bill (1 ten). When you are adding, you can **regroup** 10 ones into 1 ten. We usually call this **carrying**.

How It Works   When the digits in the ones column add up to ten or more, you need to carry one or more tens to the next column.

Example 1   What is 78 + 19?

Step 1	Step 2	Step 3
Line up the digits so that ones are under ones, tens under tens, etc.	Add the ones column. Carry if necessary.	Add the tens column, including any carried number.
$\begin{array}{r} 78 \\ +\,19 \\ \hline \end{array}$	$\begin{array}{r} \overset{1}{7}8 \\ +\,19 \\ \hline 7 \\ \uparrow \end{array}$ 8 + 9 = 17, so put the 7 in the ones place and carry the 1 to the tens place.	$\begin{array}{r} \overset{1}{7}8 \\ +\,19 \\ \hline 97 \\ \uparrow \end{array}$ 1 + 7 + 1 = 9

To understand what *carrying* or *regrouping* is all about, look at the drawing below.

Add 9 + 8, as in Example 1 on page 14:

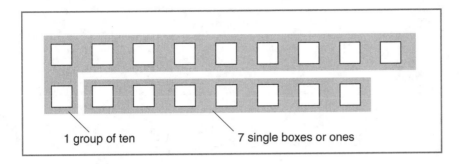

1 group of ten   7 single boxes or ones

Can you see that the one group of ten is what is "carried" to the tens column?

In some problems, you may have to carry or regroup in more than one column.

Example 2   What is the sum of 972 + 14 + 245?

Step 1	Step 2	Step 3
Line up the digits so that ones are under ones, tens under tens, etc.	Add the ones column. Carry if necessary.	Add the tens column, including any carried number. Carry again if necessary, and continue adding.
972 14 + 245	$\overset{1}{972}$ 14 + 245 ——— 1	$\overset{11}{972}$ 14 + 245 ——— 1,231

Try It   Find the following sums. Carry or regroup if necessary.

5,451 + 389	769 + 38	99 + 11

Did you get 5,840, 807, and 110 as answers? If not, go back and be sure you carried correctly.

Practice Find the following sums.
Carry if necessary.

1.	60	28	36	21	77
	+19	+12	+25	+93	+ 9

2.	590	248	3,096	219	97
	+ 39	+102	+2,509	+234	+129

3. 208 + 12 = _____   321 + 406 = _____   895 + 175 = _____

4. 157 + 92 = _____   333 + 971 = _____   450 + 375 = _____

Solve the following word problems. Use only necessary information.

5. Yoshi spent $19 on a computer printer ribbon, $16 on a lamp, and $5 on paper. He paid with a $50 bill. How much did he spend in all?

_____

6. The ticket collector at the baseball game took in 634 tickets before the third inning, and 198 tickets after that. How many total tickets did he collect?

_____

7. Last year, 12,440 people lived in Belle County. An additional 1,808 people moved there this year. What is the number of people living in Belle County this year?

_____

8. For her messenger job, a driver rode from Uptown to Delta and back again. Refer to the street map below. How many miles did she travel?

_____

Answers are on pages 115 to 116.

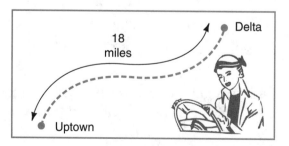

16

Subtracting Whole Numbers

What You Know If one amount is taken away from another amount, this is called **subtraction**. When you look at your paycheck, you probably see that some money has been taken out for taxes. This is an example of how subtraction plays a part in everyday life.

Knowing **basic subtraction facts** is an important skill. See how quickly you can get the answers to these subtraction problems:

10 − 3 = _____   8 − 2 = _____   9 − 7 = _____

There are different ways to "read" these subtraction problems. For the first one, you might say "ten take away three" or "ten minus three" or "three subtracted from ten." What are some different ways to read 8 − 2 and 9 − 7?

The answers to the subtraction problems above are 7, 6, and 2. Before you go on, be sure that you understand the math facts just discussed.

How It Works In subtraction problems, the order of the numbers is important. For example, 10 − 3 is *not* the same as 3 − 10. This is different from addition, in which order is not important. For example, 10 + 3 means the same as 3 + 10.

Once you are comfortable subtracting single digits, you are ready to subtract larger numbers.

Example 1   What is 43 − 21?

Step 1	Step 2	Step 3
Line up the digits. Be sure to put the number you are subtracting *from* in the top position. 43 − 21	Subtract the ones column. 43 − 21 2	Subtract the tens column. 43 − 21 22

Be careful to put the bigger number on top in a subtraction problem. The first number given in a problem is not always the one that should go on top.

Example 2   Mrs. Tran paid $35 for office supplies. She started out with $89 in her purse. How much money does she have left?

Step 1	Step 2	Step 3
Line up the digits. Be sure to put the number you are subtracting *from* in the top position. 89 − 35	Subtract the ones column. 89 − 35 4	Subtract the tens column. 89 − 35 54

Sometimes you may have a subtraction problem that involves more than two numbers. In this case, turn the problem into two separate problems.

Example 3 $989 - 54 - 824 = ?$

Step 1	Step 2	Step 3
Line up the digits of the first two numbers. Put the number you are subtracting *from* on top. $\begin{array}{r} 989 \\ -\ 54 \end{array}$	Subtract. $\begin{array}{r} 989 \\ -\ 54 \\ \hline 935 \end{array}$	Now subtract the next number from the answer in Step 2. $\begin{array}{r} 935 \\ -824 \\ \hline 111 \end{array}$

Try It   Try the following subtraction problems. Use the steps above if you need help.

What is 30 subtracted from 192?

If you line up your numbers correctly and put 192 on top, you will get the correct answer, which is 162.

What is $99 - 36 - 20$?

First find $99 - 36$. Then subtract 20. The answer is 43.

Practice   Practice finding these subtraction facts.

1. $10 - 2 =$ _____ $8 - 7\ =$ _____ $12 - 4 =$ _____ $6 - 3 =$ _____

2. $9 - 7\ =$ _____ $14 - 5 =$ _____ $8 - 3\ =$ _____ $9 - 5 =$ _____

Subtract the following.

3.
$\begin{array}{r} 38 \\ -15 \end{array}$
    $\begin{array}{r} 75 \\ -12 \end{array}$
    $\begin{array}{r} 99 \\ -\ 9 \end{array}$
    $\begin{array}{r} 42 \\ -41 \end{array}$
    $\begin{array}{r} 87 \\ -21 \end{array}$

4.
$\begin{array}{r} 34 \\ -22 \end{array}$
    $\begin{array}{r} 61 \\ -\ 1 \end{array}$
    $\begin{array}{r} 44 \\ -32 \end{array}$
    $\begin{array}{r} 75 \\ -22 \end{array}$
    $\begin{array}{r} 11 \\ -\ 9 \end{array}$

Subtract the following. Remember to turn the problem
into two separate problems.

5. 87 – 50 – 12 = _____   368 – 124 – 33 = _____

Solve the following problems. Remember to use only the
necessary numbers.

6. Maggie read 124 pages of a
244-page book. How many
more pages does she have left
to read?

_____

7. On Thursday, Anna bought
145 tropical fish for her pet
shop. By the weekend, she had
sold 33 of them, at $3 each.
How many tropical fish did
she have left?

_____

8. Peter and Crystal have been
married 141 days. If there are
365 days in a year, how many
more days remain until their
1-year anniversary?

_____

9. In 8 hours, Jack drove 420 miles.
His total trip will be 550 miles.
How many more miles will Jack
have to drive?

_____

10. Ming works as a carpenter's
assistant. She cut 32 centimeters
off the board pictured below.
How many centimeters long is
the remaining board?

_____

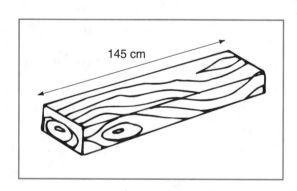

145 cm

Answers are on page 116.

Follow-Up. Check your answers to Problems 1 to 10 on page 19 and above
using a calculator.

UNIT 1   Whole Numbers

Subtracting and Borrowing

What You Know   Suppose you have a ten-dollar bill in your pocket. You owe a friend $4. How can you pay him if he does not have $6 to give you back?

A solution you probably have used in the past is to "break" the ten-dollar bill at a store or bank. In other words, you give someone the ten-dollar bill, and that person gives you 10 one-dollar bills in exchange. Then you have the correct bills to give the friend the $4 you owe him.

This regrouping or "borrowing" 10 ones to replace 1 ten is common in math problems.

How It Works   When a digit in a number being subtracted is larger than the digit in the top number, you need to **borrow** or **regroup**.

Example 1   Subtract 78 from 93.

Step 1	Step 2	Step 3
Line up the digits. Be sure to put the number you are subtracting *from* in the top position.	Subtract the ones column. Since you cannot take 8 away from 3, borrow 1 ten from the tens column and add it to the ones column.	Subtract the tens column.
$$\begin{array}{r} 93 \\ -78 \\ \hline \end{array}$$	When you borrow 1 ↙ from 9, that leaves 8. $$\begin{array}{r} {}^{8}\!\!\not{9}{}^{1}3 \\ -7\ 8 \\ \hline 5 \\ \uparrow \end{array}$$ ← Add the 10 you borrowed to the 3 ones: $10 + 3 = 13$ $13 - 8 = 5$	$$\begin{array}{r} {}^{8}\!\!\not{9}{}^{1}3 \\ -78 \\ \hline 15 \\ \uparrow \end{array}$$ $8 - 7 = 1$

In some subtraction problems, you will need to borrow more than once. Let's look at an example.

Example 2   323 − 88 = ?

Step 1	Step 2	Step 3	Step 4
Line up the digits. If you cannot subtract ones, borrow a ten. Then subtract.	Now move to the tens column. If the bottom number is larger than the top, borrow one hundred.	Subtract the tens.	Subtract the hundreds.
$$3\ \overset{1}{\cancel{2}}\ {}^{1}3 \\ -\ \ 8\ 8 \\ \hline 5$$	$$\overset{2}{\cancel{3}}\ \overset{11}{\cancel{2}}\ {}^{1}3 \\ -\ \ \ 8\ 8 \\ \hline 5$$	$$\overset{2}{\cancel{3}}\ \overset{11}{\cancel{2}}\ {}^{1}3 \\ -\ \ \ 8\ 8 \\ \hline 3\ 5$$	$$\overset{2}{\cancel{3}}\ \overset{11}{\cancel{2}}\ {}^{1}3 \\ -\ \ \ 8\ 8 \\ \hline 2\ 3\ 5$$
The 2 in the tens column becomes a 1. The borrowed 10 is added to the ones: 10 + 3 = 13.	The 3 in the hundreds column becomes a 2. The 1 in the tens column becomes 11: 10 + 1 = 11.		

When you subtract from a number with zeros in it, there is an extra step in borrowing and regrouping.

Example 3   302 − 18 = ?

Step 1	Step 2	Step 3	Step 4
Line up the digits. If you cannot subtract ones, you'll need to regroup a ten.	If there is a zero in the tens column, borrow a hundred. You now have 10 tens in the tens column.	Regroup one of the tens, leaving 9, and add 10 ones to the ones column.	Subtract the ones, tens, and hundreds.
$$302 \\ -\ 18$$	$$\overset{2}{\cancel{3}}{}^{1}0\ 2 \\ -\ \ \ 1\ 8$$	$$\overset{2}{\cancel{3}}\,\overset{9}{\cancel{1}}\overset{}{\cancel{0}}{}^{1}2 \\ -\ \ \ 1\ 8$$	$$\overset{2}{\cancel{3}}\,\overset{9}{\cancel{1}}\overset{}{\cancel{0}}{}^{1}2 \\ -\ \ \ 1\ 8 \\ \hline -2\ 8\ 4$$

Try It   Juan Carlito had $88. He bought the bag
pictured below. How much money did he have left?

Since you cannot take the 9 away from the 8 in the ones
column, you'll have to borrow:

$$\begin{array}{r} {}^{7}\cancel{8}{}^{1}8 \\ -\ 2\ 9 \\ \hline 5\ 9 \end{array}$$

905 − 436 = ?

The bottom number in the ones column is larger than the
top number. You can't regroup from the tens column, so
first regroup a hundred.

$$\begin{array}{r} {}^{8}\cancel{9}{}^{1}0\ 5 \\ -\ 4\ 3\ 6 \end{array}$$

Then borrow from the tens column, and subtract.

$$\begin{array}{r} {}^{8}\cancel{9}\ {}^{9}\cancel{0}{}^{1}5 \\ -\ 4\ 3\ 6 \\ \hline 4\ 6\ 9 \end{array}$$

Practice   Subtract. Borrow if necessary.

1. 58 − 19	175 − 146	93 − 9	42 − 28	80 − 21
2. 134 − 26	60 − 1	264 − 139	75 − 29	41 − 18
3. 2,912 − 1,573	204 − 139	405 − 316	1,039 − 981	123 − 55

Estimate an answer to the following subtraction problems. As a first step, round each number to the nearest ten or hundred. Then subtract.

4. 36 – 14 _____   14 – 9 _____   37 – 11 _____

5. 349 – 29 _____   210 – 9 _____   499 – 51 _____

Use a calculator to do these subtraction problems.

9. 2,309 – 654 = _____   19,498 – 7,211 = _____

10. 139 – 48 – 12 = _____   4,308 – 4,267 – 5 = _____

Solve these word problems. Use only necessary information.

11. From his monthly check of $1,040, Mr. Pham spends $560 for his rent and food. How much does he have left over each month?

_____

12. Cynthia poured 37 ounces from the pitcher shown below. How many ounces remained in the pitcher?

_____

64 Ounces

13. Paulo brought 50 pounds of potatoes to sell at the farmers' market. He had 23 pounds left at the end of market day. How many pounds did he sell?

_____

14. Mrs. Santiago estimated that the number of people who came to the county fair would be about 1,050. The actual number was 945. How much higher was Mrs. Santiago's estimate than the actual number of people?

_____

Answers are on pages 116 to 117.

Multiplying Whole Numbers

What You Know How many cans of soda do you get in three six-packs?

One way of writing this problem is 6 + 6 + 6. You know from addition that you get 18 cans. Another way to think about this problem is with multiplication. Three groups with 6 cans in each group is written like this:

6 × 3 = ?   or   3 × 6 = ?

The order in which numbers are written in a multiplication problem does not matter. **Multiplication** is just a faster way of adding up groups of numbers.

How It Works One of the most important sets of facts when you multiply whole numbers is the **multiplication table**. This table includes all the basic multiplication facts you'll need to solve many problems.

		2	3	4	5	6	7	8	9	10	11	12
1	1	2	3	4	5	6	7	8	9	10	11	12
2	2	4	6	8	10	12	14	16	18	20	22	24
3	3	6	9	12	15	18	21	24	27	30	33	36
4	4	8	12	16	20	24	28	32	36	40	44	48
5	5	10	15	20	25	30	35	40	45	50	55	60
6	6	12	18	24	30	36	42	48	54	60	66	72
7	7	14	21	28	35	42	49	56	63	70	77	84
8	8	16	24	32	40	48	56	64	72	80	88	96
9	9	18	27	36	45	56	63	72	81	90	99	108
10	10	20	30	40	50	60	70	80	90	100	110	120
11	11	22	33	44	55	66	77	88	99	110	121	132
12	12	24	36	48	60	72	84	96	108	120	132	144

To use a multiplication table, multiply any number in the top shaded row by any number in the left shaded column. The point at which the left column and the top row meet is your answer.

Example 1   What is 8×6?

	2	3	4	5	6	7	8	9	10	11	12
1	1	2	3	4	5	6	7	8	9	10	11	12
1	1	2	3	4	5	6	7	8	9	10	11	12
2	2	4	6	8	10	12	14	16	18	20	22	24
3	3	6	9	12	15	18	21	24	27	30	33	36
4	4	8	12	16	20	24	28	32	36	40	44	48
5	5	10	15	20	25	30	35	40	45	50	55	60
6	6	12	18	24	30	36	42	48	54	60	66	72
7	7	14	21	28	35	42	49	56	63	70	77	84
8	8	16	24	32	40	48	56	64	72	80	88	96
9	9	18	27	36	45	56	63	72	81	90	99	108
10	10	20	30	40	50	60	70	80	90	100	110	120
11	11	22	33	44	55	66	77	88	99	110	121	132
12	12	24	36	48	60	72	84	96	108	120	132	144

Find 8 in the top row. Follow the column down to 6 in the left column. The number in that square, 48, is the answer.

Try It   Use the table on page 25 to find answers to these.

$3 \times 7 =$ _____ $9 \times 5 =$ _____ $4 \times 8 =$ _____ $7 \times 6 =$ _____

Did you get 21, 45, 32, and 42? If not, review the steps for reading the multiplication table.

When you can quickly multiply single digits, you are ready for longer multiplication problems.

To multiply a number with two or more digits by a single digit, put the larger number on top, and multiply each digit by the bottom number, *always starting at the right*.

Example 2   What is 6×211?

Step 1	Step 2	Step 3
Put the larger number on top, and line up the smaller number under it.	Multiply the ones column. Put your answer in the ones place.	Multiply the tens column and then the hundreds column. Put each answer in the correct place.
211 $\times$ 6	211 $\times$ 6 6 ← $6 \times 1 = 6$	211 $\times$ 6 1,266 $6 \times 1 = 6$ $6 \times 2 = 12$

When you multiply by a two-digit number, multiply each digit to find a **partial product**. Then add the partial products to find the total answer.

Example 3   Multiply 12 by 130.

Step 1	Step 2	Step 3	Step 4
Put the larger number on top, and line up the smaller number under it.	Multiply the ones column.	Multiply the tens column. Begin this partial product *under the tens column.*	Add the partial products.
130 × 12	130 × 12 260	130 × 12 260 130 Leave this space empty, or put a zero here.	130 × 12 260 130 1,560

Try It   Try using the steps above to multiply these numbers.

41
× 9

132
× 23

Did you get 369 and 3,036? If so, good work. If not, review the steps above and try again.

Practice   Multiply by one digit.

1. 71   111   34   601   103
  × 8   × 5   × 2   × 9   × 3

2. 21   143   642   420   101
  × 7   × 2   × 2   × 4   × 8

Multiply by two digits. Remember to line up partial products correctly.

3.

11	221	34	301	143
$\times 78$	$\times\ \ 33$	$\times 22$	$\times\ \ 19$	$\times\ \ 23$

4.

12	211	44	312	103
$\times 12$	$\times\ \ 11$	$\times 22$	$\times\ \ 13$	$\times\ \ 23$

Solve these word problems. Use only necessary information.

5. Meg's car can travel 33 miles on 1 gallon of gas. How many miles can it travel on 12 gallons?

_____

7. Patrick bought 12 cases of soda and 200 hot dogs for his company cookout. Each case of soda holds 24 cans. How many cans did he buy in all?

_____

6. Each class period at the Glenview Adult Learning Center is 50 minutes long. If there are 6 periods per day, how many total minutes do students spend in class?

_____

8. Jorge's son is 5 feet tall. His daughter is 4 feet tall. There are 12 inches in 1 foot. How many inches tall is Jorge's son?

_____

Answers are on page 117.

Follow-Up. Make a set of multiplication flash cards with a partner. Put one multiplication fact on one side of each card, and put the answer on the other side of the card. Make up a card for each problem in the multiplication table. Then take turns showing each other the cards. Keep doing this until you both know all the answers.

Multiplying and Carrying

What You Know Suppose you borrowed money from three co-workers. You have to pay each $15 back. How much money will you pay back in all?

$15
× 3

When you multiply numbers like these, you need to do an extra step. You already learned about "carrying" in addition. In this lesson, you'll do the same thing with multiplication.

$15
× 3
$45

How It Works Multiplying and carrying are not difficult if you work neatly and carefully. You already know how to do all the steps; you just need to practice putting them together.

Example 1   What is 5 × 24?

Step 1	Step 2	Step 3
Put the larger number on top, and line up the smaller number under it.	Multiply the ones column. Carry to the next column if necessary.	Now multiply the tens column. Add the carried number *after* you've multiplied.
24 × 5	$\overset{2}{2}4$ × 5 —— 0 Since 4 × 5 = 20, put a 0 in the ones place and carry the 2 to the tens place.	$\overset{2}{2}4$ × 5 —— 120 5 × 2 = 10; then add the carried 2 to get 12.

As with addition, you sometimes may have to carry in more than one column. Here, it is especially important to line up your numbers carefully.

Example 2 Multiply 5 by 132.

Step 1	Step 2	Step 3	Step 4
Put the larger number on top, and line up the smaller number under it.	Multiply the ones column. Carry if necessary.	Multiply the tens column. Add any carried number. Carry to the hundreds column if necessary.	Multiply the hundreds column.
$\begin{array}{r} 132 \\ \times\ \ 5 \end{array}$	$\begin{array}{r} \overset{1}{132} \\ \times\ \ 5 \\ \hline 0 \end{array}$ $\begin{array}{l}15 \times 2 = 10,\\ \text{so carry the 1.}\end{array}$	$\begin{array}{r} \overset{11}{132} \\ \times\ \ 5 \\ \hline 60 \end{array}$ $\begin{array}{l}5 \times 3 = 15\\ 15 + 1 = 16\\ \text{Write 6 and}\\ \text{carry the 1.}\end{array}$	$\begin{array}{r} \overset{11}{132} \\ \times\ \ 5 \\ \hline 660 \end{array}$ $\begin{array}{l}5 \times 1 = 5\\ 5 + 1 = 6\end{array}$

When you multiply by more than one digit, it is even more important to carry neatly and carefully.

Example 3 $45 \times 24 = ?$

Step 1	Step 2	Step 3	Step 4
Put the larger number on top, and line up the smaller number under it.	Multiply the ones column. Carry if necessary.	Multiply the tens column. Start the partial product in the tens place.	Add the partial products.
$\begin{array}{r} 45 \\ \times 24 \end{array}$	$\begin{array}{r} \overset{2}{45} \\ \times\ 24 \\ \hline 180 \end{array}$ $\begin{array}{l}5 \times 4 = 20,\\ \text{so carry the 2.}\end{array}$ $\begin{array}{l}4 \times 4 = 16\\ 16 + 2 = 18\end{array}$	$\begin{array}{r} \overset{1}{45} \\ \times\ 24 \\ \hline 180 \\ 90 \end{array}$ $\begin{array}{l}2 \times 5 = 10,\\ \text{so carry the 1.}\\ 2 \times 4 = 8.\\ \text{Add the 1:}\\ 8 + 1 = 9\end{array}$	$\begin{array}{r} 45 \\ \times\ 24 \\ \hline 180 \\ 90 \\ \hline 1,080 \end{array}$

Try It Multiply the following numbers. Carry when necessary.

251 × 4 = _____   104 × 23 = _____

If you multiplied and carried correctly, you got 1,004 and 2,392.

When multiplication problems have several steps, rounding and estimating are useful skills. You can use estimation to check your multiplication, and you can use estimation when you do not need an exact answer.

Example 4   56 × 21 = _____?

56 rounds to ⟶ 60
21 rounds to ⟶ 20

60 × 20 = 1,200, so 56 × 21 ≈ 1,200.

Practice Multiply the following numbers. Carry if necessary.

1.
23	45	903	217
× 4	× 9	× 6	× 3

2. 107 × 5 = _____   37 × 8 = _____   50 × 7 = _____

Multiply the following numbers. Be careful to line up the partial products in the correct places.

3.
125	500	309	112	240
× 15	× 34	× 25	× 10	× 22

4.
225	210	300	512	245
× 30	× 15	× 42	× 10	× 20

5. 150 × 35 = _____   12 × 278 = _____   18 × 256 = _____

Estimate an answer to these multiplication problems. As a first step, round off one or both of the numbers.

6. 405 × 11 _____   198 × 19 _____   35 × 121 _____

Use a calculator to find the solutions to these problems.

7. 175 × 30 = _____   180 × 36 = _____   25 × 116 = _____

Solve the following word problems. Be careful to use only the necessary numbers.

8. As a salad preparer, Naomi can put together 35 chef salads in 1 hour. How many salads can she prepare in her 12-hour shift?

_____

9. A TravelEase Airways plane carries 245 passengers on its Boston–Miami flight each day. How many passengers does it carry in a 7-day week?

_____

10. Antonio's monthly car payment is $175. How much does Antonio pay in 1 year?

_____

11. A chemist pours 14 vials of liquid into a dish. The vials are the size pictured at the right. How many total milliliters did she pour into the dish?

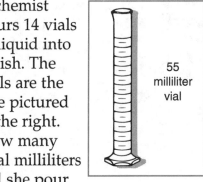

55 milliliter vial

_____

12. To find the area of a rectangle, multiply the length by the width. What is the area of the rectangle below?

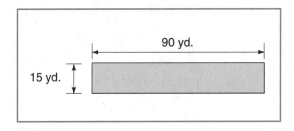

90 yd.

15 yd.

_____

Answers are on pages 118 to 119.

Dividing Whole Numbers

What You Know   Have you ever split a lunch or dinner check with friends? If so, you have used division. For example, if you and two friends eat a meal that cost $15, you divide $15 by 3, and you each pay $5.

There are different ways to write this division problem, but every division problem has three parts. They are labeled here:

$$15 \div 3 = 5$$

Dividend   Divisor   Quotient

$$5 \leftarrow \text{Quotient}$$
$$3\overline{)15}$$

Divisor   Dividend

Both statements say "15 divided by 3 equals 5."

How It Works   Dividing small numbers is easy if you know your multiplication facts. For example:

If you know that	You also know that
$5 \times 4 = 20$	$20 \div 5 = 4$ and $20 \div 4 = 5$
$3 \times 6 = 18$	$18 \div 6 = 3$ and $18 \div 3 = 6$
$8 \times 6 = 48$	$48 \div 8 = 6$ and $48 \div 6 = 8$

Can you see that division is the opposite of multiplication? In division, you take a whole amount and divide it into groups of equal size. In multiplication, you put groups of equal size together to find a total amount.

Just as multiplication of large numbers involves multiplying and adding, dividing larger numbers involves dividing and subtracting.

Example 1   What is 60 divided by 5?

Step 1	Step 2	Step 3	Step 4	Step 5
Set up the problem. Be sure that the number you are dividing into goes *inside* the division symbol. $5\overline{)60}$	Begin dividing from left to right. $1 \leftarrow$ 5 goes $5\overline{)60}$ into 6 one time	Multiply quotient by divisor, and then subtract. $\begin{array}{r} 1 \\ 5\overline{)60} \\ \underline{5} \leftarrow 5 \times 1 = 5 \\ 1 \leftarrow 6 - 5 = 1 \end{array}$	Bring down the next digit. $\begin{array}{r} 1 \\ 5\overline{)60} \\ \underline{5} \\ 10 \leftarrow \text{Bring} \\ \text{down} \\ \text{the 0.} \end{array}$	Divide again, putting a new digit in the quotient. $\begin{array}{r} {}^{10 \div 5 = 2} \\ 12 \\ 5\overline{)60} \\ \underline{5} \\ 10 \\ \underline{10} \\ 0 \end{array}$

The quotient (answer) is 12 (60 divided by 5 equals 12).

Sometimes when you are dividing, the divisor cannot be divided into the first digit of the dividend. Let's see how to handle this type of problem.

Example 2 $1228 \div 4 = ?$

Step 1	Step 2	Step 3	Step 4	Step 5
Set up the problem. $4\overline{)1,228}$	If you cannot divide into the first digit, try the first two digits. $\begin{array}{r} 3 \\ 4\overline{)1,228} \end{array}$ 4 cannot go into 1. Try 12; 4 goes into 12 three times.	Multiply quotient by divisor and then subtract. $\begin{array}{r} 3 \\ 4\overline{)1,228} \\ \underline{12} \\ 0 \end{array}$	Bring down the next digit and divide. Since 4 cannot go into 2, write a 0 above the 2 in the quotient and bring down the next digit. $\begin{array}{r} 30 \\ 4\overline{)1,228} \\ \underline{12} \\ 028 \end{array}$	Divide again, putting a new digit in the quotient. $\begin{array}{r} 307 \\ 4\overline{)1,228} \\ \underline{12} \\ 028 \\ \underline{28} \\ 0 \end{array}$

Sometimes, a number will not divide evenly into another number. In this case, there will be a **remainder** as part of the answer.

Example 3 Divide 362 by 8.

Step 1	Step 2	Step 3	Step 4
Set up the problem.	Begin dividing.	Bring down the next digit and divide.	If the division does not come out evenly, write the remainder (r) as part of the answer.

8)‾362‾	8 goes into 36 ↙ four times.
	     4
	8)‾362‾
	   32‾
	     4 ← 36 − 32 = 4
		45
		8)‾362‾
		   32‾
		   42
		   40‾
			45
			8)‾362‾
			   32‾
			   42
			   40‾
			    2
			
			Answer: 45 r2

Try It   Dividing can be tricky. The following problems have been started for you. Try to see what steps have been taken, and then complete the division.

     5   1   6
5)‾270‾   2)‾372‾   3)‾1,840‾
  25‾

Did you get these answers: 54, 186, and 613 r1?

 Math Tip Remember that a division problem that looks like this: 450 ÷ 3 can also be written like this: 3)‾450‾.

Practice   Answer these division problems and
then write a related multiplication equation.

1. 35 ÷ 7 = _____ so _____ × _____ = 35

2. 24 ÷ 3 = _____ so _____ × _____ = 24

Divide the following.

3. 5)‾1‾2‾0   3)‾3‾6   8)‾8‾1‾6   4)‾2‾1‾6

4. 155 ÷ 5 = _____   224 ÷ 4 = _____   168 ÷ 8 = _____

Some of these division problems may have a remainder.
Remember to include the remainder (r) as part of the answer.

5. 275 ÷ 8 = _____   153 ÷ 9 = _____   448 ÷ 8 = _____

Solve these word problems. Remember to use only
necessary information.

6. An extra-large pizza has 12
slices. How many slices could
each person in a four-person
family get if each eats the same
amount?

_____

8. A construction worker cut the
chain-link fencing below into
six equal sections. How many
feet was each section?

_____

7. A truck driver has to cover 966
miles in three days. How many
miles will he drive each day if
he splits the trip up evenly?

_____

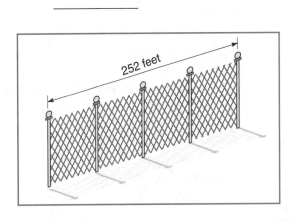

252 feet

Answers are on page 119.

Dividing Larger Numbers

What You Know   Suppose there are 120 people coming to a party. Each table can seat 10 people. How many tables are needed?

For a problem like this one, you'll need to divide a three-digit number by a two-digit number.

$$\begin{array}{r} 12 \\ 10\overline{)120} \end{array}$$

How It Works   Dividing by a two- or three-digit number is much the same as dividing by a single digit. The steps are exactly the same. You are just working with larger numbers, which can be tricky.

Example 1   Divide 364 by 14.

Step 1	Step 2	Step 3	Step 4
Set up the problem.	Begin dividing.	Subtract and bring down the next digit.	Repeat the steps to divide the next number.
$14\overline{)364}$	14 goes into 36 two times. $\begin{array}{r} 2 \\ 14\overline{)364} \\ \underline{28} \leftarrow 14 \times 2 = 28 \end{array}$	$\begin{array}{r} 2 \\ 14\overline{)364} \\ \underline{28} \\ 84 \end{array}$	$\begin{array}{r} 26 \\ 14\overline{)364} \\ \underline{28} \\ 84 \\ \underline{84} \quad 14 \times 6 = 84 \\ 0 \end{array}$

It can sometimes be tricky to figure out how many times the divisor can "go into" the dividend. Estimation is a useful skill to use.

For example, suppose you are dividing 2,205 by 63. You need to find out how many times 63 goes into 220 in order to get the first digit of the answer.

To get a good estimate, just divide the first digits in each number. For example:

$63\overline{)2{,}205}$ → 6 goes into 22 three times, so try 3.

Example 2 Divide 2,205 by 63.

Step 1	Step 2	Step 3	Step 4
Set up the problem. $63\overline{)2{,}205}$	Estimate to find how many times 63 will go into 220. $\begin{array}{r} 3 \\ 63\overline{)2{,}205} \end{array}$ Try 3 ← because 6 goes into 22 three times	Multiply, subtract, and bring down the next digit. $\begin{array}{r} 3 \\ 63\overline{)2{,}205} \\ \underline{189} \\ 315 \end{array}$	Repeat the steps to divide the next number. $\begin{array}{r} 35 \\ 63\overline{)2{,}205} \\ \underline{189} \\ 315 \\ \underline{315} \\ 0 \end{array}$ 63 goes into 315 five times, since 6 goes into 31 five times.

Try It Look at the steps that have been completed in each of these division problems. Then complete the division.

$\begin{array}{r} 8 \\ 42\overline{)3402} \end{array}$ $\begin{array}{r} 1 \\ 28\overline{)448} \end{array}$ $\begin{array}{r} 2 \\ 31\overline{)713} \end{array}$

Did you get 81, 16, and 23? You are correct if you did.

Estimating is useful when an exact answer is not needed in a division problem. By rounding off the numbers, you can get numbers that are easier to work with.

Example 3 There are 128 ounces in a gallon. About how many gallons are in the vat pictured at the right?

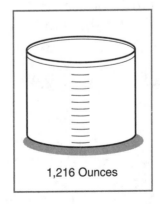

1,216 Ounces

To find an answer, you need to divide 64 into 1,216.
Round off the numbers to come up with a good estimate.

128 rounds to $\longrightarrow$ 100
1,216 rounds to $\longrightarrow$ 1,200

$1,200 \div 100 = 12$, so $1,216 \div 128 = 12$
There are about 12 gallons in the vat.

Practice Divide.

1. $20\overline{)440}$ $\qquad$ $15\overline{)375}$ $\qquad$ $38\overline{)722}$ $\qquad$ $12\overline{)144}$

2. $300 \div 15 =$ _____ $273 \div 13 =$ _____ $340 \div 17 =$ _____

Do these division problems. There may be remainders in some of them.

3. $275 \div 18 =$ _____ $140 \div 17 =$ _____ $448 \div 80 =$ _____

4. $153 \div 33 =$ _____ $360 \div 20 =$ _____ $516 \div 16 =$ _____

Solve the following division problems. Be sure to use only necessary information.

5. A chef poured 288 ounces of soup into pint containers. A pint is equal to 16 ounces. How many pints did she fill?

6. At his warehouse job, Jake can pack 24 circuit boards into a crate that holds 10 pounds. If Jake has 1,152 boards to pack, how many crates can he fill?

_____

_____

7. At a speed of 55 miles per hour, how many hours will it take to travel the route below?
(Hint: Divide the distance by the rate of speed.)

8. Raisa bought 14 candles for a total of $42. How much did each candle cost?

_____

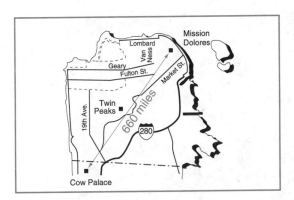

_____

Answers are on page 120.

Follow-Up. Working in groups of two or three, write your own division word problems. Use situations from your life, such as grocery shopping, budgeting, and job tasks. Exchange problems with other groups to solve.

Unit
2

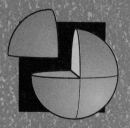

Decimals

What Are Decimals?

What You Know   We use the decimal system all the time in everyday life. In fact, it would be very difficult for any of us to survive in this world without *some* understanding of decimals. For example, if *one dollar* is expressed in numbers as $1.00, how would you write *one dollar and fifty cents*? _____

If you wrote $1.50, then you have written a decimal. Our money system is based on **decimals**.

How It Works   In Unit 1, you learned about place value and whole numbers. Now let's look at place value and numbers that are *smaller than 1*.

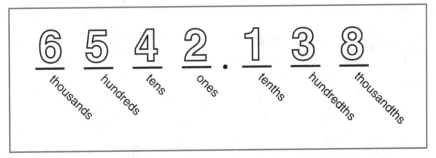

Both whole-number and decimal places extend further, but for this book and in most of everyday life, the decimal place values shown above are enough to know.

Let's take a closer look at what decimals really are.

Example 1   How much of the circle is shaded?

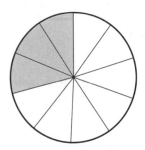

As you can see, the circle is divided into ten equal parts, or **tenths**. Three parts of the whole, or three tenths, are shaded. According to the chart, the tenths place is the first one following the decimal point. Therefore, three tenths can be written as .3.

Try It Look at the circle at the right. Shade in .7 (seven tenths) of the circle.

Did you shade in seven of the ten parts? Can you see that .7 is less than one whole?

Now think about tenths on the number line. Which is greater, .7 or .3? Use the number line below to help you.

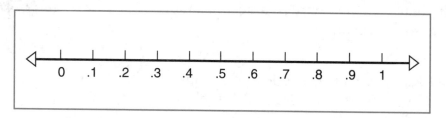

You can see that .7 (seven tenths) is closer to 1 than .3 (three tenths), so .7 is greater than .3.

Now let's look more closely at **hundredths**. The box below is divided into 100 squares: 10 rows with 10 squares in each row (10 × 10 = 100).

Example 2 How much of the box is shaded? Write the amount as a decimal.

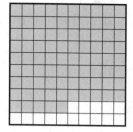

Since 85 squares out of 100 are shaded, write .85 (85 hundredths).

A **mixed decimal** is a number that contains both a whole number and a decimal.

Example 3   How many boxes are shaded?

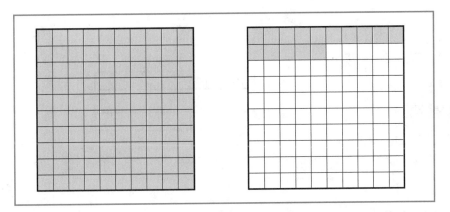

The boxes at the right are divided into 100 equal parts. One whole box is shaded, and 25 parts of the second box are shaded. We can say that *one and twenty-five hundredths* boxes are shaded.

1.25
↑
The decimal point is read as *and*.

Try It   Write a decimal that tells how many boxes are shaded.

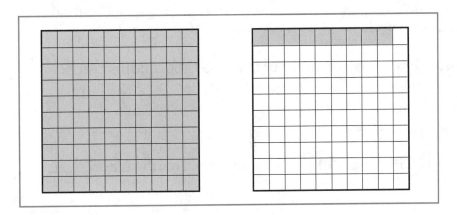

Did you write 1.9? If so, you have a good understanding of mixed decimals.

Math Tip   Remember that all the numbers that come *before* the word *and* are whole numbers. They go in front of, or to the left of, the decimal point. Any numbers *after* the word *and* are decimals and should follow the decimal point.

Practice   What part of each figure below is shaded?

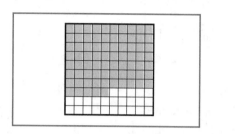

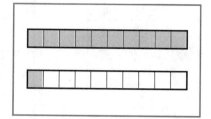

1. _____

3. _____

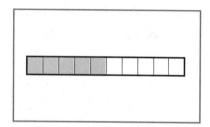

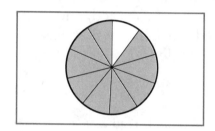

2. _____

4. _____

Use the place value chart on page 43 to answer these questions. Then write each mixed decimal in words.

5. In the number 1,298.145, what number is in the thousandths place?
_____

_____

6. In the number 421.16, what number is in the hundredths place? _____

_____

7. In the number 8,961.045, what place is the zero in? _____

_____

8. In the number 987.12, what place is the 7 in? _____

Answers are on page 121.

Zeros in Decimals

What You Know You can see the importance of zeros in the decimal system if you think about these two money amounts. Which is more?

$1.20 or $1.02

In the first number, the 2 is in the dimes, or tenths, place. In the second number, the 2 is in the pennies, or hundredths, place. The zero "holds" the 2 in the hundredths place. From your experience, you know that $1.20 is more than $1.02.

How It Works A zero may or may not change the value of a decimal. It depends on where the zero is in the number.

Example 1 Which is greater, .7 or .70?

Look at these shaded figures. Can you see that .7 = .70?

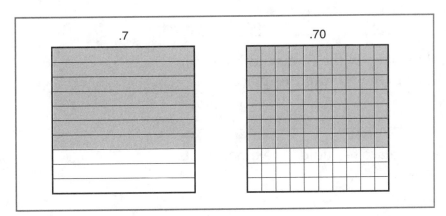

One zero or more than one zero added *after* the last digit after a decimal point will *not* change the value of the decimal.

.7 = .70 = .700

Example 2   Is .5 the same as .05?

Look at these shaded figures. You can see that five tenths (.5) is a lot larger than five hundredths (.05).

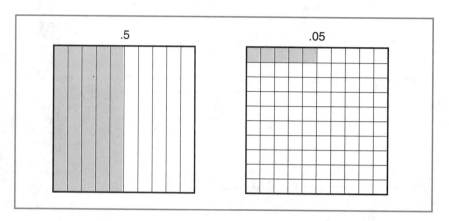

The zero in .05 "holds" the 5 in the hundredths place.

A zero added *before* the last digit after a decimal point *does change* the value of the decimal. For example, .5 ≠ .05 ≠ .005. (The symbol ≠ means "does not equal.")

When you write decimals, you may need to use one or more zeros to hold other digits in the correct place (this is called a *placeholder zero*).

Example 3   Write 23 thousandths as a decimal.

————— . ————— ————— —————

You know that the 3 must be in the thousandths place. Therefore, the 2 will be in the hundredths place:

————— . ————— 2   3

Use a zero in the tenths place to "hold" the other digits where they belong: .023.

Try It   Write the following decimals.

two and nine hundredths _____ . _____ _____

↑

Use a placeholder zero here.

seventeen and twelve thousandths

_____ _____ . _____ _____ _____

↑

Use a placeholder zero here.

six hundredths _____ . _____ _____

↑

Use a placeholder zero here.

If you wrote 2.09, 17.012, and .06, you have used placeholder zeros correctly!

Practice   Circle the correct decimal.

1. thirty-five and seventy-five thousandths	35.75	.3575	35.075
2. three hundred fifty thousandths	.350	3.050	350.001
3. ninety-eight hundredths	9,800	.098	.98
4. seven and two hundredths	.702	7.02	700.200
5. eighty-eight hundredths	8,800	88.01	.88
6. nine tenths	.9	.09	9.1

Answers are on page 121.

Rounding and Estimating Decimals

What You Know   Estimating with decimals is a lot like estimating with money — something you probably do without even thinking about it. Answer the following question as quickly as you can by using rounded numbers.

Suppose you have $50 in your bank account. You have the following bills to pay: $19.90 for telephone and $24.99 for a credit card company. Do you have enough in your account to pay these bills?

You could round each bill to the nearest dollar. The bill for $19.90 rounds to $20, and $24.99 rounds to $25. Then you can add these rounded numbers: $20 + $25 = $45. The bills add up to *approximately* $45. The account can cover all the bills.

How It Works   Estimating decimals is as easy as estimating whole numbers. You've already seen that, when you estimate with money, you can round to the nearest dollar. With other decimals, you just round to the nearest whole number. Try this:

If you round	Then you can round
$1.99 to $2	1.99 to 2
$10.11 to $10	10.11 to _____
$99.50 to $100	_____ to _____

That's right: 10.11 rounds to 10 and 99.50 rounds to 100.

Think of rounding decimals on a number line, just as you did with rounding whole numbers.

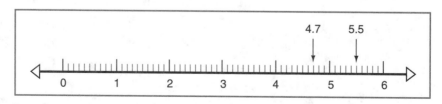

When you round a decimal, you find the whole number closest to it.

Example 1   Round 4.7 to the nearest whole number.

   Step 1   Between what two whole numbers is 4.7?
                  4.7 is between 4 and 5.

   Step 2   Is 4.7 closer to 4 or to 5?
                  You can see on the number line that 4.7 is closer to 5 than to 4.

   Step 3   Round the number.
                  4.7 ≈ 5

 Math Tip Remember: if a number is *exactly halfway* between two numbers, round to the higher number. For example, 5.5 rounds up to 6, *not* down to 5.

Sometimes, you may want to round a decimal to another decimal place instead of to a whole number. Use a number line for this kind of rounding as well.

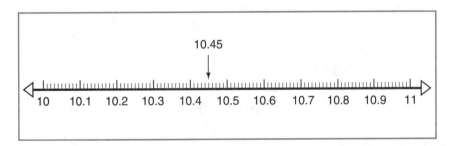

Example 2   What is 10.45 rounded to the nearest tenth?

   Step 1   Between what two tenths is 10.45?
                  Use the number line if you need to.
                  10.45 is between 10.4 and 10.5.

   Step 2   Is 10.45 closer to 10.4 or to 10.5?
                  You can see on the number line that 10.45 is exactly halfway between 10.4 and 10.5.

 Math Tip Look at the digit to the right of the place you are rounding to. If that number is 5 or more, round up.

   Step 3   Round the number.
                  10.45 ≈ 10.5

Try It Use either number line on pages 50 and 51 to help you round these decimals.

3.78 to the nearest tenth   _____

3.78 to the nearest whole number   _____

1.5 to the nearest whole number   _____

1.34 to the nearest tenth   _____

Are your answers 3.8, 4, 2, and 1.3? If so, you have a good understanding of rounding decimals.

Practice Round the following decimals to the nearest whole number.

1. 4.98 _____   125.03 _____   35.5 _____

2. 9.91 _____   80.4 _____   12.7 _____

Round these decimals to the nearest tenth.

3. 10.25 _____   3.08 _____   24.02 _____

4. 2.11 _____   6.48 _____   1.55 _____

Use rounded numbers to answer the following questions.

5. Wanda gave a clerk a $20 bill to pay for a $2.89 bottle of shampoo and a $5.12 bottle of aspirin. About how much change should Wanda receive?

_____

6. Sam wrote a $404.50 check against his bank balance of $678.98. Approximately what will his new balance be?

_____

7. Carl has about $30 in his wallet. Can he afford to pay for his $17.03 dinner as well as his date's $17.55 dinner?

_____

8. A landscaper bought 8 bushes at $24.99 each. Approximately what was her total bill?

_____

Solve the following problems using estimates.

9. The gas price listed at Bob's Garage is $1.399 per gallon. What is the price per gallon, rounded to the nearest cent, or penny?

_____

10. A deli counter worker sliced 12.78 pounds of turkey breast. How much did she cut to the nearest tenth of a pound?

_____

11. To the nearest whole pound, how much did the deli counter worker from Problem 10 slice?

_____

12. Tuan's car odometer read 24,875.4 miles. Round this figure to the nearest mile.

_____

Answers are on page 121.

Follow-Up. One very useful activity that involves estimating money amounts is writing up a household budget. Try rounding off your monthly expenses for food, clothing, shelter, and other necessities. How do these expenses compare with your income?

Adding and Subtracting Decimals

What You Know   As you've done in the previous lessons, think about decimals in terms of money. You'll see how much you already know about adding and subtracting too! Think about these two problems:

$$\begin{array}{r} 3.40 \\ +\ 2.5 \\ \hline \end{array} \qquad \begin{array}{r} 10.90 \\ -\ \ 3.5 \\ \hline \end{array}$$

If these problems look confusing at all, put a dollar sign in front of each number. Then add zeros after each 5. Look more familiar? Now can you add and subtract?

How It Works   Adding and subtracting decimals is like adding and subtracting whole numbers. Just remember to *line up the decimal points one under the other*.

Example 1   6.4 + 1.7 = ?

Step 1	Step 2	Step 3
Line up the decimal points.	Add from the right, starting with the decimals. Carry right across the decimal point.	Add the whole numbers and bring down the decimal point in the answer.
$\begin{array}{r} 6.4 \\ +\ 1.7 \\ \hline \end{array}$	$\begin{array}{r} \overset{1}{6}.4 \\ +\ 1.7 \\ \hline 1 \end{array}$	$\begin{array}{r} \overset{1}{6}.4 \\ +\ 1.7 \\ \hline 8.1 \end{array}$

When decimals have different place values, line up the decimal points, and then *add zeros after the last digit* to give all numbers the same number of places. Remember that doing this will *not* change the value of the decimal.

Example 2 Howard biked 3.75 miles on Thursday, 2.8 miles on Friday, and 4.4 miles on Saturday. What was Howard's total mileage?

Step 1	Step 2	Step 3
Line up the decimal points and add zeros.	Add from the right, starting with the decimals. Carry across the decimal point.	Add the whole numbers and bring down the decimal point to the answer.
3.75 2.80 ← Add zeros + 4.40	¹ 3.75 2.80 + 4.40 .95	¹ 3.75 2.80 + 4.40 10.95

Example 3 A carpenter started work with the dowel shown below. He sanded off .5 centimeters. How many centimeters is the sanded dowel?

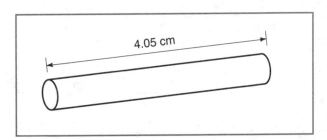

4.05 cm

Step 1	Step 2	Step 3
Line up the decimal points and add zeros if needed.	Subtract from the right, starting with the decimals. Borrow right across the decimal point.	Subtract the whole numbers and bring down the decimal point to the answer.
4.05 − .50 ← Add a zero.	$\overset{3}{\cancel{4}}{}^{1}.05$ − .50 .55	$\overset{3}{\cancel{4}}{}^{1}.05$ − .50 3 .55

Try It   Add or subtract the following decimals.
Be sure to line up the decimal points correctly.

3.5 + 9.08   = _____

10.975 − 7.5 = _____

105.65 + 35.9 = _____

5.5 − 3.75   = _____

Now check your work using a calculator. Be sure to press
the decimal point key in the correct order. Did you write
12.58, 3.475, 141.55, and 1.75?

Practice   Add or subtract the following decimals.
Check your work with a calculator.

1. $\begin{array}{r} 5.5 \\ +\ 3.8 \\ \hline \end{array}$ $\begin{array}{r} 9.9 \\ -\ 1.5 \\ \hline \end{array}$ $\begin{array}{r} 125.75 \\ +\ \ 10.5 \\ \hline \end{array}$ $\begin{array}{r} 78.2 \\ -\ \ 3 \\ \hline \end{array}$

2. $\begin{array}{r} 8.5 \\ +\ 2.5 \\ \hline \end{array}$ $\begin{array}{r} 2.1 \\ -\ 1.9 \\ \hline \end{array}$ $\begin{array}{r} 25.02 \\ +\ \ 8.9 \\ \hline \end{array}$ $\begin{array}{r} 18.2 \\ -\ \ 5.9 \\ \hline \end{array}$

3. 3.5 + 9.89 = _____   1.8 + 5.75 = _____   3.03 + 1.25 = _____

4. 10.5 − 8.25 = _____   9.4 − 2.05 = _____   5.25 − 4.1 = _____

Solve these addition and subtraction word problems.
Be careful to use only necessary information.

5. At the end of spring training, Peter Crane's batting average was .275. At the end of the regular season, his average was .355. By how much did his average increase?

6. Doug walks 1.5 miles to play tennis, another 1.9 miles around the lake, and then 3.4 miles home. What is the total distance that he walks?

_____   _____

7. Sally paid for a new watch with a $20 bill and a $10 bill. The watch cost $23.97. How much change should Sally receive?

_____

8. On Thursday, CeeCee's temperature was 101.4. By late that night, it had gone up to 102.4. Average body temperature is 98.6. How many degrees higher than average was her highest temperature?

_____

9. The Alvarez family's farm is 13.8 acres. If Mr. Alvarez purchases the 8.5-acre field next door, what will the total acreage be?

_____

10. A healthy mouse can run a maze in 5.75 seconds. A mouse injected with a flu virus takes 3.5 seconds longer to run the same maze. How many seconds does it take the second mouse?

_____

Answers are on page 122.

Follow-Up. Use the supermarket advertisements in the newspaper to make up a shopping list. Exchange lists with a partner. Then figure out the total cost of the items on the list. Suppose you started shopping with $200 in your bank account. How much would be left after you bought all the items on the list?

Multiplying Decimals

What You Know Here's another example of how what you know about money can help in your understanding of decimals.

How much is five quarters? or, to put it another way: What is 5 times $.25? _____

Now write your answer below — without the dollar sign — being sure to put the decimal point in the correct place:

.25
× 5
_____

If you wrote 1.25, you've correctly multiplied decimals!

How It Works Multiplying decimals is as easy as multiplying whole numbers. Once again, you just need to know where to put the decimal point in the answer. To figure out where to put the decimal point, count and add the digits to the right of the decimal point in each number being multiplied.

Example 1   Multiply 3.5 by .5.

Step 1	Step 2	Step 3
Multiply as you would with whole numbers.	Count and add the number of digits following the decimal points.	Starting at the right, count out the correct number of decimal places.
3.5 × .5 _____ 175	3.5 1 digit × .5 1 digit _____ 175 1 + 1 = 2 digits	3.5 1 digit × .5 1 digit _____ 1.75 2 digits

Try It   Using the steps on page 58, multiply the following numbers.

5.1	1.25	.06
× .3	× 2	× 9

Did you answer 1.53, 2.50, and .54? If not, go back and make sure that you counted the right number of decimal places.

Using zeros as placeholders is an important skill in multiplying decimals.

Example 2   What is .2 × .3?

Step 1	Step 2	Step 3
Multiply as you would with whole numbers.	Count the number of digits following the decimal points.	Place the decimal point in the answer, using zeros if necessary.
.2 × .3 6	.2 1 digit × .3 1 digit 6 1 + 1 = 2 digits	.2 1 digit × .3 1 digit .06 2 digits ↑ Add this zero so that the correct number of digits will follow the decimal.

Try It   The numbers below have already been multiplied for you. Put the decimal point in the correct place. Add zeros when they are needed.

3.08	.02	1.5	.125
× .2	× .01	× 1.5	× .3
616	2	225	375

Did you answer .616, .0002, 2.25, and .0375?

Multiply these decimals.

2.25	.07	3.2	5.8
× 1.5	× .05	× 1.8	× .9

Did you get 3.375, .0035, 5.76, and 5.22?

Practice   Multiply these decimals.

1. 100.25   75.9   9.9   .04
        ×   3.2   × .02   × 1.1   × .2

2. 5.5 × 1.9 = _____   90.9 × .5 = _____   16.4 × 4.4 = _____

Estimate an answer to these problems by rounding off
one or both decimals.

3. 99.9 × 1.1 _____   25.5 × 3.01 _____   4.8 × 9.7 _____

4. 9.2 × 200.1 _____   29.5 × 6.99 _____   3.25 × 4.8 _____

Use a calculator to solve these problems.

5. 3.025 × 4.9 = _____   85.5 × 1.5 = _____   .125 × 375 = _____

6. 2.85 × 6.07 = _____   1.254 × 10 = _____   .009 × .5 = _____

Solve these multiplication problems. Use only the
information needed.

7. Eddie bought 3 donuts at $.49
each. How much did he pay for
the donuts before tax?

9. A store sells 1 yard of fabric
for $1.80. Shanisha bought
2.5 yards. How much did she
pay in all?

_____

_____

8. A truck can carry 10.5 tons of
topsoil. If the truck made 12
full deliveries, how many tons
did it carry in all?

10. Carlos loaded 35 crates onto
a shelf. Each crate weighed
1.75 kilograms. What was the
total weight of the crates?

_____

_____

11. Area is found by multiplying length by width. What is the area of the figure below?

_____

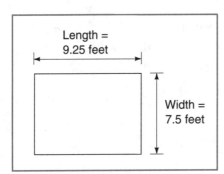

Length =
9.25 feet

Width =
7.5 feet

12. Lih earns time and a half (×1.5) for any overtime hours she works. If her regular pay is $5.50 per hour, what is her pay per hour of overtime?

_____

Answers are on pages 122 to 123.

Follow-Up. Use your calculator to make a discovery. Multiply:

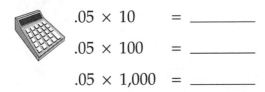

.05 × 10   = _____

.05 × 100   = _____

.05 × 1,000   = _____

What did you learn about multiplying by 10, 100, and 1,000? Apply what you learned about moving the decimal point to these problems:

.025 × 10   = _____

.025 × 100   = _____

.025 × 1,000 = _____

LESSON 16   Multiplying Decimals

Dividing Decimals by a Whole Number

What You Know Can you split a $40.50 gas bill evenly between two people? What would each person pay? _____ Put your answer in the space below.

$$2\overline{)40.50}$$

Dividing decimals is like dividing whole numbers. You just need to know where to put the decimal point.

How It Works As with all your work with decimals, the key to dividing correctly is *knowing where to put the decimal point*.

When you divide a decimal by a whole number, placing the decimal point is easy. Just put it directly above the decimal point in the dividend.

← Quotient (answer)

$$2\overline{)10.8}$$

↑
Divisor   Dividend

Example 1   What is 10.8 ÷ 2?

Step 1	Step 2	Step 3
Set up the division. $2\overline{)10.8}$	Bring the decimal point *directly* above the dividend. $2\overline{)10.8}$ ← Bring the decimal point up.	Divide as you would with whole numbers. $\begin{array}{r} 5.4 \\ 2\overline{)10.8} \\ \underline{10} \\ 0\,8 \\ \underline{8} \\ 0 \end{array}$

Sometimes a division problem isn't finished, but you run out of digits to bring down. You can add one or more zeros after the last digit to the right of the decimal point in order to finish dividing. Adding this zero does not change the value of the number. For example, 2.5 = 2.50.

Example 2   A carpenter cuts the board at the right into two equal pieces. How long is each piece?

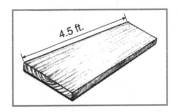
4.5 ft.

Step 1	Step 2	Step 3
Set up the division and bring the decimal point *directly* above.	Divide as you would with whole numbers.	Continue dividing, adding zeros if needed.

Step 1:
```
     .       ← Bring the
2)4.5          decimal
               point up.
```

Step 2:
```
   2.2
2)4.5
  4
  ‾
   5
```

Step 3:
```
   2.25
2)4.50    ↖ Add a zero.
  4
  ‾
   5        Bring down a
   4        zero.
   ‾      ↙
   10
   10
   ‾
    0
```

Try It   Divide the following numbers. Be sure to bring the decimal point directly up.

3)12.42   8)64.8   5)20.6

The answers are 4.13, 8.1, and 4.12.

Remember to use zeros as placeholders, just as you do with whole numbers.

Math Tip   A division problem written as 3.5 ÷ 5 can be rewritten using the symbol)‾ . So 3.5 ÷ 5 is the same as 5)3.5.

Example 3   What is $31.92 \div 3$?

Step 1	Step 2	Step 3
Set up the division and bring the decimal point *directly* above.	Divide as you would with whole numbers. Use a zero as a placeholder when necessary.	Continue dividing, adding zeros if needed.

Step 1:

$$3\overline{)31.92} \quad \leftarrow \text{Bring the decimal point up.}$$

Step 2:

$$\begin{array}{r} 10 \\ 3\overline{)31.92} \\ \underline{3} \\ 019 \end{array}$$

Since 3 does not go into 1, put a 0 above the 1. Then divide 3 into 19 instead.

Step 3:

$$\begin{array}{r} 10.64 \\ 3\overline{)31.92} \\ \underline{3} \\ 019 \\ \underline{18} \\ 12 \\ \underline{12} \\ 0 \end{array}$$

Try It   Divide these numbers. Use zeros as placeholders.

$$12\overline{)24.36} \qquad 9\overline{).018} \qquad 3\overline{)2.124}$$

Did you get 2.03, .002, and .708? If you missed a zero or put it in the wrong place, go back and review Example 3.

Practice   Divide these numbers. Use zeros correctly.

1. $5\overline{)12.505}$ $\qquad 10\overline{)3.04}$ $\qquad 4\overline{).008}$

2. $3\overline{)2.112}$ $\qquad 12\overline{)1.236}$ $\qquad 6\overline{)5.76}$

3. $301.010 \div 10 =$ _____ $.279 \div 3 =$ _____ $1.8 \div 5 =$ _____

Estimate an answer to these problems by rounding one or both numbers.

4. $50.9 \div 3.1$ _____ $8.01 \div 1.9$ _____ $9.98 \div 4.8$ _____

Solve the following word problems.

5. To build a pen for his dog, Harry bought 102 meters of wire and divided it into lengths of 25.5 meters each. How many lengths did he cut?

_____

6. Andy and Jeff split the cost of two CDs. One cost $11.40, and the other cost $14.40. How much money did each friend spend? (Hint: Add the cost of both CDs and then divide by 2.)

_____

7. Four friends split the cost of a retirement gift for a coworker. The gift cost $36.08. How much should each person pay?

_____

8. A jeweler cut the chain shown below into three equal pieces. How long will each piece be?

_____

25.2 inches

9. A chef's helper divided 25.45 grams of spice into five jars. How many grams were in each jar?

_____

10. Tishan bought three shirts that cost $14.50, $18.00, and $10.40. What was the average price of the shirts? (Hint: An average cost is found by adding the total cost and dividing by the number of items.)

_____

Answers are on pages 123 to 124.

Follow-Up. Use your calculator to divide 100 by 3. What do you get?

This kind of number is called a **repeating decimal**. In a repeating decimal, one digit or a sequence of digits keeps repeating forever.

To work with a repeating decimal, we usually round it off to the nearest hundredth. 0.333333333 = 0.33. Divide these numbers on your calculator, and then round to the nearest hundredth.

200 ÷ 3 _____   12.5 ÷ 3 _____   47 ÷ 9 _____

Dividing a Decimal by a Decimal

What You Know Suppose you have $1.50. You want to buy some candy bars that cost $.25 each. How many can you buy?

You probably can figure this out by just picturing the coins in your head, or by counting how many quarters there are in $1.50. If you wanted to write a division problem to figure this out, here's how it would look:

$$.25\overline{)1.50} \quad 6$$

In other words, you'd be dividing a decimal by a decimal. This lesson will show you how.

How It Works When you divide *by* a decimal — in other words when the divisor is a decimal — your first step is to make the divisor a whole number. You can do this by moving the decimal point to the right. However, you must also move the decimal point *the same number of places* in the dividend.

This is also true when you divide by a mixed decimal. You learned about mixed decimals in Lesson 12. They are numbers that contain a whole number and a decimal.

When you divide by a mixed decimal, you must make the divisor a whole number. You do this by moving the decimal point to the right. Then you must move the decimal point in the dividend the same number of places to the right.

Example 1 Divide 20.52 by .4.

Step 1	Step 2	Step 3
Set up the division. How many places to the right do you move the decimal point to make the divisor a whole number? .4) 20.52 Move the decimal point one place to the right: .4 → 4	Move the decimal point the same number of places in *both divisor and dividend.* .4.) 20.5.2 Move the decimal point one place to the right in both numbers.	Now put the decimal point directly above in the answer, and divide. $$\begin{array}{r} 51.3 \\ 4.\overline{)205.2} \\ \underline{20} \\ 5 \\ \underline{4} \\ 12 \\ \underline{12} \\ 0 \end{array}$$

Sometimes when you are dividing a decimal by a decimal, you'll need to add a zero.

Example 2 What is 1.3 ÷ .65?

Step 1	Step 2	Step 3
Set up the division. How many places to the right do you move the decimal point to make the divisor a whole number? .65) 1.3 Move the decimal point two places to the right: .65 → 65	Move the decimal point the same number of places in both divisor and dividend. *Add a zero to the dividend if necessary.* .65.) 1.30. Add a zero. Move the decimal point two places to the right.	Now put the decimal point directly above in the answer, and divide. $$\begin{array}{r} 2 \\ 65.\overline{)130.} \\ \underline{130} \\ 0 \end{array}$$ The answer is 2. You can drop the decimal point because the answer doesn't have a decimal point.

Try It   Move the decimal points the correct number of places.

$.01 \overline{)7}$ $\qquad$ $2.8 \overline{)21.84}$ $3 \overline{)9.075}$

Do your problems now look like this? If so, divide.

$01 \overline{)700.}$ $28 \overline{)218.4}$ $3 \overline{)9.075}$

Your answers should be: 700, 7.8, and 3.025.

Practice   Divide these numbers.

1. $.25 \overline{)5.750}$ $\qquad$ $1.4 \overline{)28.42}$ $\qquad$ $.035 \overline{)70}$

2. $22.5 \div .5 =$ _____ $.009 \div .03 =$ _____ $75.75 \div 2.5 =$ _____

Use a calculator to divide the following numbers.

3. $.34 \div .2 \;\;=$ _____ $.06 \div .15 =$ _____ $75 \div 12.5 =$ _____

4. $9000 \div .9 =$ _____ $4.1 \div 4 \;\;=$ _____ $.055 \div .11 =$ _____

Solve the following problems. Be careful to use only necessary numbers.

5. Su Lin's gross pay last week was $266.25. If she worked 35.5 hours, what is her hourly wage?

6. A chef's assistant paid $12.46 for a crate of parsley that cost $.89 per bunch. How many bunches were in the crate?

_____

_____

7. Danita drove 307.8 miles on 10.8 gallons of gas. How many miles per gallon of gas did she get? (Hint: To find gas mileage, divide the miles driven by the number of gallons of gas.)

_____

8. How many $.15 apples can you buy with $6?

_____

Answers are on page 124.

Unit
3

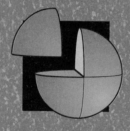

Fractions

Parts of a Whole

What You Know Fractions are a part of your life. Circle any fraction you are familiar with.

Cooking measurement $\frac{1}{3}$ cup oil

Mileage distance $\frac{3}{10}$ of a mile

Linear measurement $\frac{1}{2}$ inch

Each of these examples expresses a value smaller than a whole. A **fraction** is a part of a whole.

How It Works You already know that a decimal is part of a whole. In the case of decimals, the whole is divided into 10, 100, or 1,000 parts. A fraction can be a part of a whole that has been divided into *any* number of parts. The bottom number of a fraction, called the **denominator**, is the total number of parts that the whole is divided into. The top number of a fraction, called the **numerator**, is the actual number of parts you have.

Example 1 What fraction of the circle is shaded?

——— ← Numerator
——— ← Denominator

You can see that the circle is divided into three equal parts. Therefore, 3 is the denominator. Two of the parts are shaded. Therefore, 2 is the numerator. So $\frac{2}{3}$ of the circle is shaded.

Example 2   What part of the figure below is shaded? Write your answer as a fraction and as a decimal.

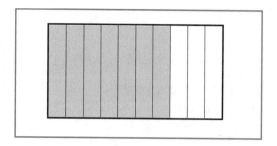

Fraction   Decimal

‾‾‾‾‾‾   ←— Numerator   ‾‾‾ . ‾‾‾
         ←— Denominator

The rectangle is divided into 10 equal parts, and 7 of the parts are shaded. Did you write $\frac{7}{10}$ and .7?

Try It   Suppose there are 8 workers on the night shift at Swanheart Plastics Factory. Of these workers, 5 are women. What fraction of the workers is female?

‾‾‾‾‾‾   ←— Numerator
         ←— Denominator

The whole is 8; the part is 5. You should have written $\frac{5}{8}$.

Practice   What fraction of each figure is shaded?

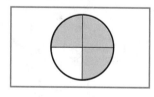   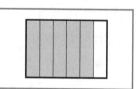   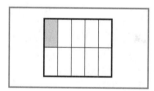

1. ‾‾‾‾‾‾‾‾‾   ‾‾‾‾‾‾‾‾‾   ‾‾‾‾‾‾‾‾‾

   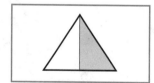   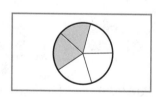

2. ‾‾‾‾‾‾‾‾‾   ‾‾‾‾‾‾‾‾‾   ‾‾‾‾‾‾‾‾‾

Write a fraction for each part of a whole described below.

3. A yard is equal to 36 inches. Marita was 19 inches at birth. What fraction of a yard was she?

_____

4. Gabriel was present for 4 out of 5 workdays last week. What fraction of the workdays was he at work?

_____

5. Tom's son counted 43 pennies in his piggy bank. What fraction of a dollar do these pennies represent?

_____

6. Three out of every ten customers at Chih's Variety Store buy lottery tickets. What fraction of Chih's customers buy tickets?

_____

7. There were 105 people riding the G subway downtown. At the Gateway stop, 31 people got off. What fraction of the riders got off at Gateway?

_____

8. Francesca spent 13 minutes on a customer service call. What fraction of an hour did she spend?

_____

Answers are on page 125.

Follow-Up. Cut out a strip of paper like the one below. You can use this piece of paper to get an idea of the size of different fractions.

Fold the paper in half. At the fold mark, write $\frac{1}{2}$.

Unfold the paper and then fold it again into thirds. Mark each fold $\frac{1}{3}$.

Continue folding the strip into different fractions and label each fold. Then compare the different sizes.

Equivalent Fractions

What You Know   In the figure below, shade $\frac{1}{2}$ of the first box. Then shade $\frac{2}{4}$ of the second box. Which shaded portion is larger?

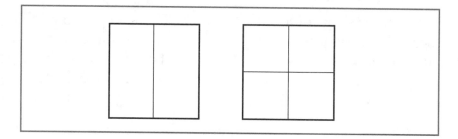

Can you see that the shaded parts of each box are *the same size*? Write the fractions on the lines below.

_____ = _____

You have just learned that $\frac{1}{2}$ and $\frac{2}{4}$ are **equivalent**, or **equal, fractions**. Even though the numbers in $\frac{2}{4}$ are bigger, the fraction $\frac{2}{4}$ stands for the same space in the box as $\frac{1}{2}$.

How It Works   To find an equivalent fraction, multiply both denominator and numerator by the *same* number. You can also divide both numerator and denominator by the same number.

You can do this without changing the value of the original fraction. Here's why:

A fraction with the same numerator and denominator is equal to 1.

$$\frac{3}{3} = 1 \qquad \frac{25}{25} = 1 \qquad \frac{102}{102} = 1 \qquad \frac{5}{5} = 1$$

When you multiply a number by 1, the answer you get is the same number:

$$3 \times 1 = 3 \quad 10 \times 1 = 10 \quad 35 \times 1 = 35 \quad 29 \times 1 = 29$$

Therefore, when you multiply a fraction by a fraction with the same numerator and denominator, the *value* of the fraction stays the same even though the numbers are bigger.

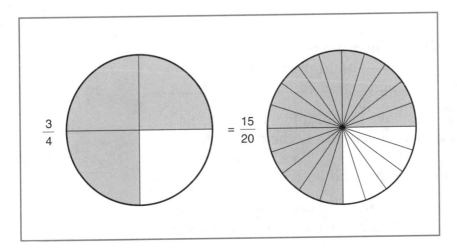

Example 1   Find a fraction equivalent to $\frac{3}{4}$.

$\frac{3}{4} \times \frac{5}{5} = ?$

This is a way of multiplying by 1.

$\frac{3}{4} \times 1 = \frac{3}{4}$

Multiplying by 1 doesn't change the value of the fraction.

$\frac{3}{4} \times \frac{5}{5} = \frac{15}{20}$

The numbers are bigger but the value of the fraction is the same.

There are many other fractions that are equivalent to $\frac{3}{4}$. You just have to multiply both numerator and denominator by *the same number*.

Often in your work with fractions, you will need to find an equivalent fraction with a given denominator. In other words, you won't be trying to find just any equivalent fraction, as you were in Example 1.

Example 2 $\frac{3}{5} = \frac{?}{10}$

In this example, you are given the new denominator. Ask yourself, "What was done to the first denominator to get the second one?" You know that $5 \times 2 = 10$, so multiply the numerator by 2 as well.

$\frac{3}{5} \times \frac{2}{2} = \frac{6}{10}$

Try It Find the equivalent fractions. In the first two, you can use any new denominator. In the second two, you are given the denominator.

$$\frac{4}{5} = \frac{?}{?} \qquad \frac{3}{7} = \frac{?}{?} \qquad \frac{5}{6} = \frac{?}{30} \qquad \frac{7}{8} = \frac{?}{16}$$

For the first two fractions, there are many possible answers. Just be sure that you multiplied both numerator and denominator by the same number.

For the fraction $\frac{5}{6}$, you have to multiply 6 by 5 to get a denominator of 30. Therefore, multiply the numerator 5 by the same number, and you get $\frac{25}{30}$. For the fraction $\frac{7}{8}$, you need to multiply both top and bottom by 2: $\frac{14}{16}$.

When you *divide* both numerator and denominator by the same number, you are reducing the fraction to **lower terms**.

Example 3 $\qquad \frac{8}{12} = \frac{?}{3}$

You know that to get a 3 as a new denominator, you must divide 12 by 4. Therefore, divide 8 by 4 as well.

$$\frac{8}{12} \div \frac{4}{4} = \frac{2}{3}$$

Since you can't divide the numerator and denominator of $\frac{2}{3}$ evenly by the same number, $\frac{2}{3}$ is said to be in **lowest terms**.

Try It Find equivalent fractions for these fractions by either *multiplying* or *dividing* the numerator and denominator by the same number.

$$\frac{4}{12} = \frac{?}{3} \qquad \frac{2}{7} = \frac{8}{?} \qquad \frac{15}{20} = \frac{?}{4}$$

Did you do the following?

$$\frac{4}{12} \div \frac{4}{4} = \frac{1}{3} \qquad \frac{2}{7} \times \frac{4}{4} = \frac{8}{28} \qquad \frac{15}{20} \div \frac{5}{5} = \frac{3}{4}$$

Practice Find the number that makes each pair of fractions equivalent.

1. $\dfrac{3}{5} = \dfrac{9}{}$ $\dfrac{1}{8} = \dfrac{}{24}$ $\dfrac{12}{15} = \dfrac{4}{}$

2. $\dfrac{6}{8} = \dfrac{3}{}$ $\dfrac{2}{3} = \dfrac{}{9}$ $\dfrac{24}{48} = \dfrac{}{12}$

3. $\dfrac{1}{6} = \dfrac{5}{}$ $\dfrac{4}{7} = \dfrac{12}{}$ $\dfrac{1}{8} = \dfrac{}{32}$

4. $\dfrac{9}{18} = \dfrac{1}{}$ $\dfrac{3}{4} = \dfrac{}{40}$ $\dfrac{5}{9} = \dfrac{15}{}$

Find equivalent fractions in these word problems.

5. A salad dressing recipe calls for $\frac{3}{4}$ of a cup of oil. Tess's measuring cup is divided into eighths. How many eighths should she put in?

_____

6. Jason rollerbladed $\frac{8}{10}$ of a mile home from work. How many fifths of a mile did he skate?

_____

7. The fish Michel caught on his company outing was $\frac{12}{16}$ of a pound. How many fourths of a pound was the fish?

_____

8. The shelf Betty is building is $\frac{2}{3}$ of a foot wide. How many twelfths of a foot wide is the shelf?

_____

Answers are on page 125.

Follow-Up. With a partner, see how many equivalent fractions for the fractions below you can write in 3 minutes. Remember that you can multiply or divide by the same number.

$\dfrac{5}{20}$ $\dfrac{8}{40}$ $\dfrac{10}{100}$ $\dfrac{2}{4}$

Rounding and Estimating

What You Know   The box at the right is divided into 100 equal squares. Shade in 99 squares of the box. What fraction of the box is shaded?

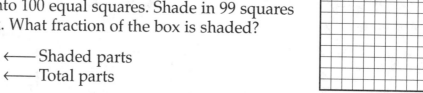

_____ ←—— Shaded parts
←—— Total parts

Estimate how much of the box is shaded.
Circle your answer below.

Not even close to   About $\frac{1}{2}$ the box   Almost the whole box
the whole box

How It Works   You may have noticed that the number of parts shaded — 99 — was very close to the total parts — 100. *When the numerator and denominator are very close in value, the fraction is close to 1, or the whole.*

Now shade in three squares of the box at the right (which is the same as the box above). Is the fraction shaded not close to the whole box, about $\frac{1}{2}$ of the box, or almost the whole box?

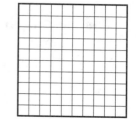

You can probably see that the shaded portion is much, much less than the whole. *The greater the difference between the numerator and the denominator, the smaller the fraction.*

It's also easy to tell whether a fraction is close to $\frac{1}{2}$. Learn this rule: *If the numerator is about half the denominator, the fraction is close to $\frac{1}{2}$.* Also, if the numerator is *exactly* half the denominator, the fraction is *equal* to $\frac{1}{2}$.

Try It

$$\frac{8}{9} \qquad \frac{2}{5} \qquad \frac{1}{12} \qquad \frac{4}{9} \qquad \frac{11}{13} \qquad \frac{8}{16} \qquad \frac{9}{1000}$$

Which of the fractions above are close to $\frac{1}{2}$? _____

Which are close to 1? _____

Which are a lot smaller than 1? _____

Is there a fraction equal to $\frac{1}{2}$? _____

Did you find that $\frac{2}{5}$ and $\frac{4}{9}$ are close to $\frac{1}{2}$ because their numerators are close to half the denominators?

The fractions $\frac{8}{9}$ and $\frac{11}{13}$ are pretty close to 1 because the numerators and denominators are close in value.

The fractions $\frac{1}{12}$ and $\frac{9}{1000}$ have numerators that are a lot smaller than their denominators, so these fractions are a lot smaller than 1.

In the fraction $\frac{8}{16}$, the numerator is exactly half the denominator, so $\frac{8}{16}$ is equal to $\frac{1}{2}$.

Practice

$$\frac{1}{8} \qquad \frac{7}{14} \qquad \frac{2}{21} \qquad \frac{5}{6} \qquad \frac{9}{11} \qquad \frac{10}{21} \qquad \frac{12}{13}$$

1. Which fractions above are equal to $\frac{1}{2}$? _____

2. Which fractions are a lot smaller than 1? _____

3. Which fractions are close to 1? _____

4. Which fractions are close to, but not equal to, $\frac{1}{2}$? _____

Add a numerator or a denominator to make these fractions close to 1.

5. $\dfrac{8}{} \qquad \dfrac{10}{} \qquad \dfrac{}{10} \qquad \dfrac{6}{} \qquad \dfrac{}{5} \qquad \dfrac{9}{} \qquad \dfrac{}{4}$

Add a numerator or a denominator to make these fractions close to $\frac{1}{2}$ or equal to $\frac{1}{2}$.

6. $\dfrac{3}{\ \ }$　　$\dfrac{5}{10}$　　$\dfrac{2}{8}$　　$\dfrac{9}{9}$

Add a numerator or a denominator to make these fractions a lot smaller than 1.

7. $\dfrac{3}{\ \ }$　　$\dfrac{9}{100}$　　$\dfrac{2}{85}$　　$\dfrac{6}{13}$

Answer the following problems.

8. As an administrative assistant, Jeannette is responsible for filing all data disks. She has filed $\frac{4}{5}$ of the disks. Is she just starting the job, has she finished about half the job, or is she almost finished?

_____

9. Yolanda has read 11 pages of a 130-page book. Has she read a small fraction of the book, about half the book, or almost the whole book?

_____

10. An overnight delivery employee has driven 86 miles on the road from Ashland to Town Crossing. Has he traveled a large fraction of his trip, about half the trip, or a small fraction of the trip?

_____

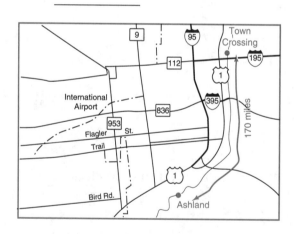

Answers are on page 126.

Common Denominators

What You Know   It can be difficult adding and subtracting things that are not alike. Working with unlike things is easier if you can find what they have in common. For example, if you want to add 6 apples and 4 oranges, you might say you have 10 pieces of fruit.

Add the following. Be sure to include a label with your answer.

$$\begin{array}{r} 25 \text{ women} \\ + 20 \text{ men} \\ \hline \end{array} \qquad \begin{array}{r} 100 \text{ Chevrolets} \\ + 250 \text{ Fords} \\ \hline \end{array} \qquad \begin{array}{r} 9 \text{ Red Sox players} \\ + 9 \text{ Yankee players} \\ \hline \end{array}$$

Did you write 45 people, 350 cars, and 18 baseball players?

In this lesson, you'll learn about finding **common denominators** for fractions in order to make them easier to work with.

How It Works   To find a common denominator for two fractions, use one of these methods:

If you can divide the smaller denominator evenly into the larger denominator, use the larger denominator.

Example 1   Find a common denominator for $\frac{1}{2}$ and $\frac{5}{6}$.

Since 2 can be divided evenly into 6, 6 can be the common denominator.

$$\frac{1}{2} = \frac{3}{6} \qquad \frac{5}{6} = \frac{5}{6}$$

$$\frac{1}{2} = \frac{3}{6} \qquad \text{and} \qquad \frac{5}{6} = \frac{5}{6}$$

If the smaller denominator *cannot* be divided evenly into the larger one, go through the multiplication table of the larger denominator until you find a number that the smaller denominator divides evenly into.

Example 2   Find a common denominator for $\frac{2}{5}$ and $\frac{1}{3}$.

Since 3 cannot be divided evenly into 5, look at the multiples of 5 until you find one that 3 can divide evenly into.

$5 \longrightarrow$ no

$10 \longrightarrow$ no

$15 \longrightarrow$ Yes, because $15 \div 3 = 5$. Therefore, 15 is the common denominator.

 Math Tip You can always find a common denominator by multiplying the two denominators together. $3 \times 5 = 15$

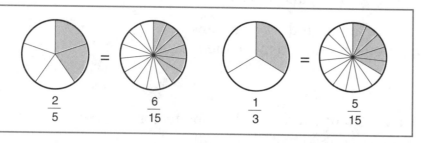

$$2 \times 3 = 6 \qquad\qquad 1 \times 5 = 5$$

$$\frac{2}{5} = \frac{6}{15} \qquad \text{and} \qquad \frac{1}{3} = \frac{5}{15}$$

$$5 \times 3 = 15 \qquad\qquad 3 \times 5 = 15$$

Try It Find a common denominator for each pair of fractions.

$\frac{3}{4}$ and $\frac{2}{3}$ \qquad $\frac{1}{8}$ and $\frac{1}{2}$ \qquad $\frac{5}{6}$ and $\frac{1}{4}$

_____ \qquad _____ \qquad _____

Did you find 12, 8, and 12 as the common denominators? If so, you are correct!

Practice   First find the common denominator for each pair of fractions. Then write equivalent fractions with the new denominator.

1. $\dfrac{4}{5}$ and $\dfrac{1}{8}$ $\dfrac{1}{10}$ and $\dfrac{3}{5}$ $\dfrac{3}{7}$ and $\dfrac{1}{4}$

$\dfrac{4}{5} = -$ $\dfrac{1}{10} = -$ $\dfrac{3}{7} = -$

$\dfrac{1}{8} = -$ $\dfrac{3}{5} = -$ $\dfrac{1}{4} = -$

2. $\dfrac{1}{6}$ and $\dfrac{1}{8}$ $\dfrac{2}{3}$ and $\dfrac{3}{4}$ $\dfrac{5}{8}$ and $\dfrac{1}{16}$

$\dfrac{1}{6} = -$ $\dfrac{2}{3} = -$ $\dfrac{5}{8} = -$

$\dfrac{1}{8} = -$ $\dfrac{3}{4} = -$ $\dfrac{1}{16} = -$

Answers are on page 126.

Adding Fractions

What You Know Use the picture at the right
to help answer this question.

What is $\frac{1}{2} + \frac{1}{2}$? _____

You can probably see that $\frac{1}{2} + \frac{1}{2} = \frac{2}{2}$,
or one whole.

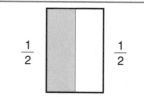

How It Works Adding fractions with *like*
(same) *denominators* is easy. You just add the numerators,
and keep the same denominator. Sometimes, when you add
like fractions, you have to *reduce* the answer to *lowest terms*.

Example 1 $\frac{1}{8} + \frac{5}{8} = ?$

Step 1	Step 2	Step 3
Add the numerators.	Keep the denominators.	Reduce the fraction to lowest terms
$\begin{array}{r} \frac{1}{8} \\ + \frac{5}{8} \\ \hline 6 \end{array}$	$\begin{array}{r} \frac{1}{8} \\ + \frac{5}{8} \\ \hline \frac{6}{8} \end{array}$	$\frac{6}{8} \div \frac{2}{2} = \frac{3}{4}$

Try It Add these fractions. Reduce to lowest terms
if necessary.

$\frac{3}{8} + \frac{4}{8} =$ ____ $\frac{1}{3} + \frac{1}{3} =$ ____ $\frac{1}{4} + \frac{1}{4} =$ ____

Did you write $\frac{7}{8}$, $\frac{2}{3}$, and $\frac{1}{2}$? If so, you're ready to go on
to adding with unlike denominators. If you had any
trouble, go back and review the steps in this lesson.

When you have to add fractions that do not have the same denominator, your first step is to write an equivalent fraction that *does* have the same denominator. Use either method you learned in Lesson 22.

Example 2   Add $\frac{1}{3} + \frac{1}{6}$.

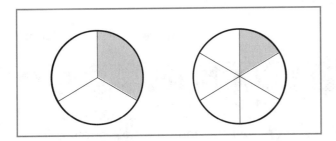

Step 1	Step 2	Step 3	Step 4
Find a common denominator. Since 3 can divide evenly into 6, use 6 as the common denominator. $\frac{1}{3} \times \frac{2}{2} = \frac{2}{6}$	Add the numerators. $\begin{array}{r} \frac{2}{6} \\ + \frac{1}{6} \\ \hline \frac{3}{} \end{array} \cdot$	Use the same denominator. $\begin{array}{r} \frac{2}{6} \\ + \frac{1}{6} \\ \hline \frac{3}{6} \end{array}$	Reduce to lowest terms if necessary. $\frac{3}{6} \div \frac{3}{3} = \frac{1}{2}$

Try It

Add $\frac{3}{8} + \frac{1}{6}$.

Step 1 $\frac{3}{8} = \frac{9}{24}$ $\frac{1}{6} = \frac{4}{24}$

Step 2 $\frac{9}{24} + \frac{4}{24} = \frac{}{24}$

Add $\frac{1}{9} + \frac{1}{3}$.

Step 1 Use 9 as a common denominator. $\frac{1}{3} = \frac{?}{9}$

Step 2 $\frac{1}{9} + \frac{}{9} = \frac{}{9}$

Did you write $\frac{13}{24}$ and $\frac{4}{9}$? If so, you are correct.

Sometimes when you add fractions, you may get a fraction that is equal to or larger than 1. In other words, the numerator may be the same size as, or larger than, the denominator. This is called an **improper fraction**. You will have to know how to change this kind of fraction to a whole number or a mixed number. A **mixed number** is a number that contains both a whole number and a fraction.

Example 3 $\frac{2}{3} + \frac{3}{4} = ?$

Step 1	Step 2	Step 3	Step 4
Find a common denominator. You can multiply 3×4 to get 12. $$\frac{2}{3} \times \frac{4}{4} = \frac{8}{12}$$ $$+ \frac{3}{4} \times \frac{3}{3} = \frac{9}{12}$$	Add the numerators. $$\frac{8}{12}$$ $$+ \frac{9}{12}$$ $$\frac{17}{12}$$ The numerator is larger than the denominator.	If the result is an improper fraction, divide the denominator into the numerator. $$12\overline{)17}$$ $$\frac{12}{5}$$	The quotient is the whole number. Put the remainder over the denominator. $$\frac{17}{12} = 1\frac{5}{12}$$

Try It

Add $\frac{4}{5} + \frac{3}{4}$. Change improper fractions to mixed numbers.

Step 1   Find a common denominator:

$$4 \times 5 = 20 \qquad \frac{4}{5} = \frac{16}{20} \qquad \frac{3}{4} = \frac{15}{20}$$

Step 2   Add: $\frac{16}{20} + \frac{15}{20} = \frac{31}{20}$

Step 3   Change the improper fraction to a mixed number:

$$\frac{31}{20} = \underline{\hspace{2cm}}$$

You should have gotten the mixed number $1\frac{11}{20}$.

Practice

Add the fractions below. Reduce if necessary. Change improper fractions to whole or mixed numbers.

1. $\frac{1}{3} + \frac{2}{3} =$ _____ $\frac{3}{8} + \frac{1}{8} =$ _____ $\frac{1}{5} + \frac{3}{5} =$ _____

2. $\frac{1}{2} + \frac{2}{3} =$ _____ $\frac{3}{8} + \frac{1}{16} =$ _____ $\frac{6}{7} + \frac{1}{4} =$ _____

3. $\frac{3}{4} + \frac{1}{3} =$ _____ $\frac{1}{10} + \frac{4}{5} =$ _____ $\frac{1}{9} + \frac{1}{3} =$ _____

Solve the following word problems. Be careful to use only the necessary numbers.

4. As a tailor's helper, Cal has to cut $\frac{1}{3}$ yard of denim and $\frac{1}{4}$ yard of nylon. How much fabric will he cut in all?

_____

5. Anthony weighed $\frac{3}{4}$ pound of melon and $\frac{5}{8}$ pound of pineapple. How many pounds of fruit did he weigh in all?

_____

6. On his newspaper route, Wayne walks $\frac{3}{5}$ of a mile and then drives $\frac{9}{10}$ of a mile. How many miles is his route altogether?

_____

7. Claudia drilled a hole in a piece of Plexiglass — the size shown below. Her supervisor asked her to widen it by $\frac{1}{8}$ centimeter. How wide will the new hole be?

_____

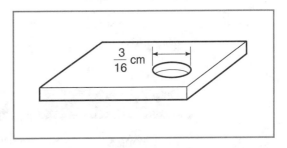

8. Mrs. Yasui filed $\frac{1}{4}$ of the office folders. Her coworker also filed $\frac{1}{4}$ of the folders. What fraction of the folders has been filed?

_____

Answers are on pages 126 to 127.

Follow-Up.

Adding fractions is a good skill to learn "hands-on." For example, take some square sheets of paper of the same size. Then cut them into halves, thirds, quarters, eighths, tenths — any fraction. Put the pieces of each square back together. Ask a partner to add them up, to see how many equal parts make up the whole square.

Subtracting Fractions

What You Know Based on what you just learned about adding fractions, see whether you can do the following subtraction problem.

$$\frac{3}{4} - \frac{1}{4} = ?$$

If you knew to subtract just the numerators and use the same denominator, you would have written $\frac{2}{4}$. Then perhaps you reduced that to $\frac{1}{2}$.

How It Works Subtracting fractions is similar to adding them. First make sure that the denominators are the same. Then subtract the numerators and keep the same denominator.

Example 1 What is $\frac{2}{3} - \frac{1}{4}$?

Step 1	Step 2	Step 3	Step 4
Change one or both denominators to get equivalent fractions.	Subtract the numerators.	Use the same denominator.	Reduce to lowest terms if necessary.
$\frac{2}{3} \times \frac{4}{4} = \frac{8}{12}$ $\frac{1}{4} \times \frac{3}{3} = \frac{3}{12}$	$\frac{8}{12}$ $-\frac{3}{12}$ $\frac{5}{}$	$\frac{8}{12}$ $-\frac{3}{12}$ $\frac{5}{12}$	$\frac{5}{12}$ is in lowest terms.

Try It Subtract these fractions. Reduce if necessary.

$$\frac{5}{8} - \frac{3}{8} = \underline{\quad} \qquad \frac{1}{3} - \frac{1}{9} = \underline{\quad} \qquad \frac{5}{6} - \frac{1}{4} = \underline{\quad}$$

Did you write $\frac{1}{4}$, $\frac{2}{9}$, and $\frac{7}{12}$?

Sometimes you may need to subtract a fraction from a whole number. In this case, you will have to borrow or rename a 1 from the whole number and change it to a fraction. Renaming a 1 is easy. Just remember that any number that is the same in both the numerator and the denominator is equal to 1.

$$1 = \frac{2}{2} = \frac{3}{3} = \frac{4}{4} = \frac{5}{5}, \text{ and so on.}$$

Example 2 What is $1 - \frac{5}{8}$?

Step 1	Step 2	Step 3	Step 4
Set up the problem.	Write 1 as a fraction using the denominator from the other fraction.	Subtract numerators as usual.	Reduce to lowest terms if necessary.
$\begin{array}{r} 1 \\ -\ \frac{5}{8} \\ \hline \end{array}$	$1 = \frac{8}{8}$ $\begin{array}{r} \frac{8}{8} \\ -\ \frac{5}{8} \\ \hline \end{array}$	$\begin{array}{r} \frac{8}{8} \\ -\ \frac{5}{8} \\ \hline \frac{3}{8} \end{array}$	$\frac{3}{8}$ is in lowest terms

Try It Rename the 1 in each problem below so that it will be easy to subtract.

$$1 - \frac{3}{4} = \underline{\quad} - \frac{3}{4} \qquad 1 - \frac{6}{7} = \underline{\quad} - \frac{6}{7} \qquad 1 - \frac{1}{5} = \underline{\quad} - \frac{1}{5}$$

You should have used $\frac{4}{4}$, $\frac{7}{7}$ and $\frac{5}{5}$ for 1.

Subtract these fractions from whole numbers. Remember that your first step should be to rename the 1 to a fraction with the same numerator and denominator.

$$1 - \frac{1}{3} = \underline{\hspace{1cm}} \qquad 1 - \frac{2}{9} = \underline{\hspace{1cm}} \qquad 1 - \frac{5}{7} = \underline{\hspace{1cm}}$$

The answers are $\frac{2}{3}$, $\frac{7}{9}$, and $\frac{2}{7}$. Did you rename the 1 correctly for each?

Practice   Subtract the following fractions. Reduce if necessary.

1. $\dfrac{7}{9} - \dfrac{3}{9} = \underline{\hspace{1cm}}$ $\qquad$ $\dfrac{11}{12} - \dfrac{1}{12} = \underline{\hspace{1cm}}$ $\qquad$ $\dfrac{5}{6} - \dfrac{1}{6} = \underline{\hspace{1cm}}$

2. $\dfrac{7}{8} - \dfrac{1}{4} = \underline{\hspace{1cm}}$ $\qquad$ $\dfrac{2}{3} - \dfrac{1}{6} = \underline{\hspace{1cm}}$ $\qquad$ $\dfrac{7}{12} - \dfrac{1}{3} = \underline{\hspace{1cm}}$

3. $\dfrac{4}{5} - \dfrac{3}{10} = \underline{\hspace{1cm}}$ $\qquad$ $\dfrac{13}{14} - \dfrac{3}{7} = \underline{\hspace{1cm}}$ $\qquad$ $\dfrac{1}{3} - \dfrac{2}{9} = \underline{\hspace{1cm}}$

4. $\begin{array}{r} 1 \\ -\ \frac{4}{5} \\ \hline \end{array}$ $\qquad\qquad$ $\begin{array}{r} 1 \\ -\ \frac{1}{6} \\ \hline \end{array}$ $\qquad\qquad$ $\begin{array}{r} 1 \\ -\ \frac{3}{4} \\ \hline \end{array}$

Solve these word problems. Use only the necessary information and reduce fractions when possible.

5. Tatiana poured $\frac{1}{3}$ cup of oil from a container that held $\frac{3}{4}$ cup. How much oil remained in the container?

6. From Tom's house to school is $\frac{9}{10}$ of a mile. From Tom's house to his son's child care center is $\frac{1}{2}$ mile along the same road. How far is it from the child care center to the school?

_____ $\qquad\qquad\qquad$ _____

7. From a 1-pound bag of nails, Hector used $\frac{5}{8}$ of a pound. What fraction of a pound was left?

_____

9. A customer service employee spent $\frac{3}{4}$ of her lunch hour helping with an account. What fraction of her 1-hour lunch does she have left?

_____

8. A sheet metal worker sanded $\frac{1}{15}$ inch from the piece of steel pictured below. How thick was the newly sanded sheet?

_____

10. A farm worker mowed $\frac{3}{8}$ acre of hay. If the entire field is $\frac{3}{4}$ acre, how much of the field did not get mowed?

_____

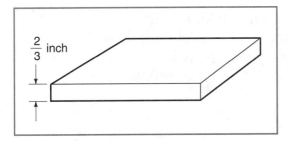

$\frac{2}{3}$ inch

Answers are on pages 127 to 129.

Follow-Up. Use a dollar bill and some coins to see whether you can make a connection between subtracting fractions and subtracting decimals. Write and solve a fraction problem for each of the following money problems. The first one is done for you.

$1.00 − $.75 = ? $.70 − $.50 = ? $.50 − $.25 = ?

$1 - \dfrac{3}{4} = \dfrac{1}{4}$ _____   _____

Adding Mixed Numbers

What You Know   Suppose the distance from your home to the grocery store is $2\frac{1}{3}$ miles. If you go another $1\frac{1}{3}$ miles, you'll get to your workplace. Can you figure out how far it is from your home to your place of work?

Perhaps just from your own experience, you can see that $2\frac{1}{3} + 1\frac{1}{3} = 3\frac{2}{3}$ miles. This kind of number is called a **mixed number** because it contains both a whole number and a fraction.

How It Works   You already know the hardest part of working with mixed numbers — adding fractions. When you add mixed numbers, you first work with the fraction part. Then you work with the whole number part.

Example 1   Add $3\frac{1}{2}$ and $6\frac{2}{3}$.

Step 1	Step 2	Step 3	Step 4	Step 5
Find a common denominator.	Add the fractions.	Add the whole numbers.	If the fraction is an improper fraction, change it to a mixed number.	Add the mixed number to the whole number part of the answer.
$3\frac{1}{2} = 3\frac{3}{6}$ $+6\frac{2}{3} = 6\frac{4}{6}$ Use 6 because it is the smallest number that both 2 and 3 can divide evenly into.	$3\frac{3}{6}$ $+6\frac{4}{6}$ $\overline{\frac{7}{6}}$	$3\frac{3}{6}$ $+6\frac{4}{6}$ $\overline{9\frac{7}{6}}$	$\frac{7}{6} = 1\frac{1}{6}$	$9 + 1\frac{1}{6} = 10\frac{1}{6}$

Try It Solve the following problem, following the steps in Example 1.

Manuel jogged $3\frac{1}{2}$ miles around the community center track. He then ran an additional $5\frac{5}{6}$ miles. How many miles did Manuel run all together?

Did you use the following steps correctly?

Step 1 Find a common denominator:
$$3\frac{1}{2} = 3\frac{3}{6} \qquad 5\frac{5}{6} = 5\frac{5}{6}$$

Step 2 Add the fractions:
$$3\frac{3}{6}$$
$$+5\frac{5}{6}$$
$$\frac{8}{6}$$

Step 3 Add the whole numbers:
$$3\frac{3}{6}$$
$$+5\frac{5}{6}$$
$$8\frac{8}{6}$$

Step 4 Change the improper fraction to a mixed number: $\frac{8}{6} = 1\frac{2}{6} = 1\frac{1}{3}$

Step 5 Add the whole and the mixed numbers:
$8 + 1\frac{1}{3} = 9\frac{1}{3}$

Estimating with mixed numbers is a useful skill. When you estimate, round mixed numbers to the nearest whole number. You'll need to use the skills you learned when you worked on estimating the size of fractions.

Try It
Estimate an answer to this problem and then find an exact answer using the examples above: $5\frac{7}{8} + 4\frac{1}{4} = ?$

Estimate:

$$5\frac{7}{8} \approx 6$$
$$\underline{+4\frac{1}{4} \approx 4}$$
$$10$$

$$5\frac{7}{8} + 4\frac{1}{4} \approx 10$$

Exact:

$$5\frac{7}{8} = 5\frac{7}{8}$$
$$\underline{+4\frac{1}{4} \quad +4\frac{2}{8}}$$
$$9 + 1\frac{1}{8} = 10\frac{1}{8}$$

Practice
Add these mixed numbers.

1. $1\frac{1}{2} + 9\frac{1}{2} =$ _____ $2\frac{2}{3} + 1\frac{1}{8} =$ _____ $5\frac{3}{4} + \frac{1}{5} =$ _____

2. $3\frac{1}{2} + \frac{1}{8} =$ _____ $8\frac{1}{8} + \frac{3}{4} =$ _____ $2\frac{1}{2} + 2\frac{1}{2} =$ _____

Estimate an answer to these problems by rounding to the nearest whole number.

3. $5\frac{1}{10} + 2\frac{4}{5} \approx$ _____ $4\frac{7}{8} + 1\frac{9}{10} \approx$ _____ $2\frac{1}{8} + 2\frac{1}{7} \approx$ _____

Solve these word problems.

4. Raymond used the road map below to travel from Bendon to Sagamore to Trenton. How many miles in all did he travel?

5. Perimeter is the distance around a figure. To find perimeter, add up the lengths of each side. What is the perimeter of the figure below?

_____

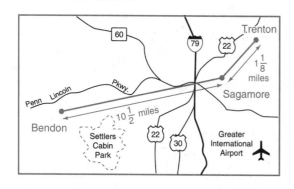

_____

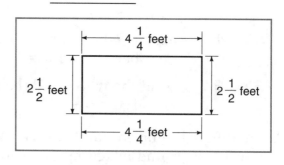

Answers are on pages 123 to 130.

Subtracting Mixed Numbers

What You Know Suppose you have three dollar bills and three quarters in your pocket. You owe your friend $2.50. What do you give your friend? How much do you have left?

From experience, you know that you should give your friend two quarters (which equals a half dollar) and two dollar bills. You have one dollar and one quarter left.

When you subtract money amounts using bills and coins, you are subtracting mixed numbers. This is how the problem above looks using mixed numbers.

$$3\frac{3}{4} = 3\frac{3}{4}$$
$$-2\frac{1}{2} = 2\frac{2}{4}$$
$$\rule{3cm}{0.4pt}$$
$$1\frac{1}{4}$$

How It Works To subtract mixed numbers, first find a common denominator. Then subtract the fractions. Finally subtract the whole numbers. Reduce if necessary.

Example 1 Subtract $2\frac{1}{4}$ from $9\frac{2}{3}$.

Step 1	Step 2	Step 3	Step 4
Find a common denominator.	Subtract the fractions.	Subtract the whole numbers.	Reduce if necessary.
$9\frac{2}{3} = 9\frac{8}{12}$ $-2\frac{1}{4} = 2\frac{3}{12}$	$9\frac{8}{12}$ $-2\frac{3}{12}$ $\rule{1.5cm}{0.4pt}$ $\frac{5}{12}$	$9\frac{8}{12}$ $-2\frac{3}{12}$ $\rule{1.5cm}{0.4pt}$ $7\frac{5}{12}$	$7\frac{5}{12}$ is in lowest terms.

When you subtract mixed numbers, you may need to borrow from the whole number and rename the 1 as a fraction. This is similar to when you renamed the 1 and subtracted a fraction from it (for example, $1 - \frac{3}{8} = \frac{8}{8} - \frac{3}{8}$).

Example 2 Rename $4\frac{1}{2}$.

You can rename 4 as $3\frac{2}{2}$ because $\frac{2}{2} = 1$.

$$4\frac{1}{2} = 3\frac{2}{2} + \frac{1}{2} = 3\frac{3}{2}$$

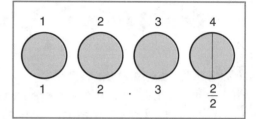

Example 3 What is $3\frac{1}{2} - 1\frac{3}{4}$?

Step 1	Step 2	Step 3	Step 4
Find a common denominator.	Borrow 1 from the whole number, and rename it using the common denominator.	Subtract the fractions.	Subtract the whole numbers.
$3\frac{1}{2} = 3\frac{2}{4}$ $-1\frac{3}{4} = 1\frac{3}{4}$	$\overset{2}{\cancel{3}}\frac{2}{4} + \frac{4}{4} = 2\frac{6}{4}$ $-1\frac{3}{4}$	$2\frac{6}{4}$ $-1\frac{3}{4}$ $\overline{\frac{3}{4}}$	$2\frac{6}{4}$ $-1\frac{3}{4}$ $\overline{1\frac{3}{4}}$

Try It Subtract the following mixed numbers. Use the steps above as a guide.

$$7\frac{1}{8}$$
$$-5\frac{1}{2}$$

Did you rewrite the problem this way?

$$\begin{array}{c} 7\frac{1}{8} \\ -5\frac{1}{2} \end{array} = \begin{array}{c} 7\frac{1}{8} \\ -5\frac{4}{8} \end{array} = \begin{array}{c} \overset{6}{\cancel{7}}\frac{1}{8} + \frac{8}{8} \\ -5\frac{4}{8} \end{array} = \begin{array}{c} 6\frac{9}{8} \\ -5\frac{4}{8} \\ \hline 1\frac{5}{8} \end{array}$$

96

Practice Subtract these mixed numbers.
Some may require renaming.

1. $7\frac{1}{8}$ $9\frac{9}{10}$ $6\frac{1}{3}$

    $-2\frac{1}{4}$ $-3\frac{3}{5}$ $-\frac{4}{9}$

2. $3\frac{1}{2} - \frac{1}{8} =$ _____ $8\frac{1}{8} - \frac{3}{4} =$ _____ $3\frac{1}{2} - 2\frac{3}{4} =$ _____

Estimate an answer to these problems by rounding to the
nearest whole number.

3. $2\frac{9}{10} - 2\frac{1}{5} \approx$ _____ $8\frac{3}{4} - 1\frac{1}{12} \approx$ _____ $20\frac{1}{8} - 5\frac{1}{9} \approx$ _____

Solve the following word problems.

4. On Friday, Reyco Stock was
selling at $51\frac{3}{4}$ points per
share. On Monday morning, the
same stock was selling at $47\frac{1}{8}$.
By how many points did the
stock go down?

_____

5. When Sheena drove her new
car out of the car lot, the car
had $374\frac{9}{10}$ miles on the
odometer. At the end of one
week, the car had $445\frac{1}{10}$ miles
on it. How many miles did
Sheena drive that week?

_____

6. Terry poured $3\frac{3}{4}$ gallons of
gasoline from the can pictured
below. How much gas was left
in the can?

_____

$10\frac{1}{2}$ Gallons

7. Brian is $48\frac{3}{4}$ inches tall. His
friend Ian is $50\frac{1}{8}$ inches. How
many inches taller is Ian than
Brian?

_____

Answers are on pages 130 to 131.

Follow-Up. Use a ruler, yardstick, or tape measure to find the perimeter of
different objects or rooms. Then use these measurements to practice working
with mixed numbers.

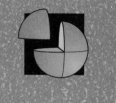

Multiplying Fractions

What You Know Think of the parts of a dollar in terms of coins:

What is half of a half-dollar? _____

Did you say a *quarter*, or a *fourth of*, a dollar?

Here is the same relationship expressed as fractions:

$$\frac{1}{2} \times \frac{1}{2} = \frac{1}{4}$$

When you need to find a fraction of a number, you multiply.

How It Works In many fraction problems, when you see the word *of*, you should multiply. Multiplying fractions is easy — you simply multiply the numerators. Then multiply the denominators.

Example 1 Franco bought $\frac{2}{3}$ of a pound of nails for a home improvement job. Half of the nails were 3 inches long. How many pounds of nails were 3 inches long?

Step 1	Step 2	Step 3
Multiply the numerators.	Multiply the denominators.	Reduce the fraction to lowest terms if necessary.
$\frac{2}{3} \times \frac{1}{2} = \frac{2}{}$	$\frac{2}{3} \times \frac{1}{2} = \frac{2}{6}$	$\frac{2}{6} = \frac{1}{3}$ $\frac{1}{3}$ pound of 3-inch nails

There is a shortcut you can use when multiplying some fractions. This shortcut is called **canceling**. If one numerator and one denominator can be divided evenly by the same number, do this division before you multiply.

Example 2 $\quad \dfrac{3}{4} \times \dfrac{2}{5} = ?$

Step 1	Step 2	Step 3
Cancel if possible. The denominator 4 and the numerator 2 can both be divided evenly by 2. $\dfrac{3}{4} \times \dfrac{2}{5} = ?$	Multiply the numerators and the denominators. $\dfrac{3}{\overset{}{\underset{2}{4}}} \times \dfrac{\overset{1}{2}}{5} = \dfrac{3}{10}$	Reduce the fraction to lowest terms if necessary. $\dfrac{3}{10}$ is in lowest terms

When you multiply whole or mixed numbers, change them into improper fractions first.

A whole number can be put in fraction form by simply putting the number over 1 (using 1 as the denominator).

$$3 = \dfrac{3}{1} \qquad 44 = \dfrac{44}{1} \qquad 12 = \dfrac{12}{1}$$

A mixed number can be written as a fraction by multiplying the whole number by the denominator and adding the numerator to this amount. Then put this number over the original denominator.

$$4\dfrac{2}{3} = 4 \times 3 + 2$$
$$= 12 + 2 = 14$$
$$2\dfrac{2}{3} = \dfrac{14}{3}$$

Example 3   What is $2\frac{1}{2} \times 4$?

Step 1	Step 2	Step 3	Step 4
Change any mixed or whole number to an improper fraction. $2\frac{1}{2} =$ $2 \times 2 = 4 + 1 = 5$ $2\frac{1}{2} = \frac{5}{2}$ $4 = \frac{4}{1}$	Cancel if possible. $\frac{5}{\underset{1}{2}} \times \frac{\overset{2}{4}}{1} = ?$	Multiply the numerators and denominators. $\frac{5}{\underset{1}{2}} \times \frac{\overset{2}{4}}{1} = \frac{10}{1}$	Write any improper fraction as a mixed number. $\frac{10}{1} = 10 \div 1 = 10$

Try It   Multiply these whole and mixed numbers. As a first step, write them as improper fractions.

$4\frac{3}{4} \times 8 = ?$

Change the numbers to improper fractions:

$4\frac{3}{4} =$ _____ $8 =$ _____

Multiply the fractions, using canceling if possible:

_____ $\times$ _____ $=$ _____

Reduce if necessary. _____

Did you come up with 38 as your answer? If not, review the steps above and try again.

Practice   Multiply these fractions, using cancelation if possible.

1. $\frac{1}{2} \times \frac{1}{4} =$ _____ $\frac{3}{8} \times \frac{2}{5} =$ _____ $\frac{5}{6} \times \frac{3}{5} =$ _____

2. $\frac{2}{3} \times \frac{7}{8} =$ _____ $\frac{3}{10} \times \frac{5}{8} =$ _____ $\frac{4}{9} \times \frac{3}{4} =$ _____

Multiply the following fractions, whole numbers, and mixed numbers. Reduce all to lowest terms.

3. $3\frac{1}{4} \times \frac{1}{2} =$ _____ $1\frac{1}{8} \times \frac{3}{4} =$ _____ $5 \times 3\frac{1}{10} =$ _____

4. $\frac{5}{6} \times 1\frac{1}{3} =$ _____ $\frac{9}{10} \times 1\frac{3}{5} =$ _____ $\frac{2}{3} \times 1\frac{1}{8} =$ _____

Solve the following word problems.

5. A recipe for soup calls for $1\frac{3}{4}$ cup of chicken broth. Ricardo wanted to make just half the recipe. How many cups of broth should he use?

_____

6. Pauline's delivery route is $10\frac{3}{4}$ miles long. If she covers this route six times each day, how many miles does she cover?

_____

7. Eumi, a plumber's assistant, fits together 12 lengths of pipe. Each length is $4\frac{1}{2}$ feet long. How many feet long is the newly fitted pipe?

_____

8. The area of a room is found by multiplying the length by the width. What is the area of the room pictured below? [Hint: Your answer will be in square yards.]

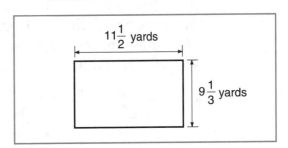

$11\frac{1}{2}$ yards

$9\frac{1}{3}$ yards

_____

Answers are on pages 131 to 133.

Follow-Up. Use a ruler, yardstick, or tape measure to find the length and width of any square or rectangular object or room. Multiply the length by the width to find the area.

Dividing Fractions

What You Know   Look at the circle below.
One half of it is shaded.

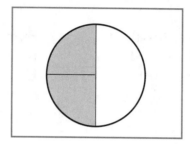

How many fourths are in this shaded half? _____

Another way to express this problem is by division of fractions.

$$\frac{1}{2} \div \frac{1}{4} = \underline{\hspace{2cm}}$$

If you wrote 2 in each blank, you're correct!

How It Works   Dividing fractions is not difficult. In fact, it is a lot like multiplying fractions, with one added step — *invert* the fraction you are dividing *by* and change the division sign to a multiplication sign. To *invert* means to flip the fraction upside down — to write the numerator as the denominator and the denominator as the numerator.

$$\frac{3}{4} \text{ inverts to } \frac{4}{3}$$

$$\frac{1}{8} \text{ inverts to } \frac{8}{1}$$

Example 1 $\dfrac{2}{3} \div \dfrac{1}{2} = \;?$

Step 1	Step 2	Step 3
Invert the divisor and change the ÷ sign to the × sign.	Multiply as usual.	Change any improper fraction to a mixed number and reduce if necessary.
$\dfrac{2}{3} \div \dfrac{1}{2} = \;?$ $\dfrac{2}{3} \times \dfrac{2}{1} = \;?$	$\dfrac{2}{3} \times \dfrac{2}{1} = \dfrac{4}{3}$	$\dfrac{4}{3} = 1\dfrac{1}{3}$

Try It   Divide these fractions using the steps above.

$$\dfrac{3}{4} \div \dfrac{1}{3} = \underline{\hspace{2cm}} \qquad\qquad \dfrac{3}{5} \div \dfrac{1}{4} = \underline{\hspace{2cm}}$$

Did you write $2\dfrac{1}{4}$ and $2\dfrac{2}{5}$?

When you divide mixed numbers, follow the same sequence of steps. First, change whole and mixed numbers to improper fractions, as you do in multiplication.

Example 2 A landscape worker has a roll of sod (grass growing in dirt) that is $6\frac{1}{4}$ meters in length. He plans to cut the sod into strips that are $1\frac{1}{4}$ meters long. How many strips of sod will he have?

Step 1	Step 2	Step 3	Step 4
Change any mixed or whole number to an improper fraction.	Invert the divisor and change the ÷ sign to the × sign.	Cancel if possible and multiply.	Write any improper fraction as a mixed or whole number.
$6\frac{1}{4} \div 1\frac{1}{4} =$ $\frac{25}{4} \div \frac{5}{4} =$	$\frac{25}{4} \div \frac{5}{4} = ?$ $\frac{25}{4} \times \frac{4}{5} = ?$	$\overset{5}{\cancel{25}}\over 4}} \times \frac{\overset{1}{4}}{\underset{1}{\cancel{5}}} = \frac{5}{1}$	$\frac{5}{1} = 5$ strips of sod

Try It Divide these fractions, following the steps given above.

$$\frac{1}{3} \div 2 = \underline{\quad} \qquad 8\frac{1}{4} \div \frac{1}{2} = \underline{\quad} \qquad 3\frac{3}{4} \div \frac{1}{4} = \underline{\quad}$$

Did you write $\frac{1}{6}$, $16\frac{1}{2}$, and 15? If you did not, review the steps above and try again.

Practice Divide these fractions.

1. $\frac{3}{4} \div \frac{1}{2} = \underline{\quad}$ $\qquad \frac{5}{6} \div \frac{5}{8} = \underline{\quad}$ $\qquad \frac{5}{8} \div \frac{1}{4} = \underline{\quad}$

2. $4 \div \frac{1}{2} = \underline{\quad}$ $\qquad 5 \div \frac{1}{8} = \underline{\quad}$ $\qquad \frac{5}{8} \div 3 = \underline{\quad}$

3. $2\frac{3}{4} \div 2 = \underline{\quad}$ $\qquad 1\frac{5}{6} \div \frac{1}{3} = \underline{\quad}$ $\qquad 2\frac{3}{5} \div \frac{2}{5} = \underline{\quad}$

4. $1\frac{7}{8} \div \frac{1}{4} = \underline{\quad}$ $\qquad 4\frac{1}{6} \div 2\frac{1}{3} = \underline{\quad}$ $\qquad 10 \div \frac{1}{10} = \underline{\quad}$

Solve these word problems.

5. The Chung family decided to break up a $10\frac{1}{2}$-mile bike ride into $5\frac{1}{4}$-mile segments. How many segments will the Chungs' trip be?

_____

6. Bob chopped down the tree pictured below, and cut it into four equal pieces. What is the length of each piece?

_____

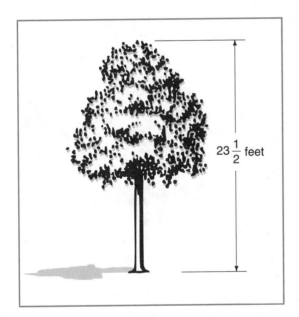

23$\frac{1}{2}$ feet

7. The pipe pictured below will be divided into five equal lengths. How long will each length be?

_____

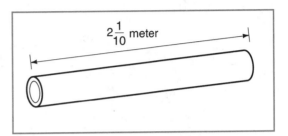

$2\frac{1}{10}$ meter

8. A butcher has 16 pounds of beef that she wants to cut into $\frac{3}{4}$-pound steaks. How many steaks will she cut?

_____

9. Each service call that Julio makes to fix a broken copy machine requires about $1\frac{1}{2}$ hours. If Julio's workday is 9 hours, how many service calls can he make in one day?

_____

Answers are on pages 133 to 136.

Follow-Up. Use your calculator to change a fraction to a decimal. All you need to do is divide the denominator into the numerator. For example, to change $\frac{3}{4}$ to a decimal, press these keys:

| C | | 3 | | ÷ | | 4 | | = |

Unit Reviews

Unit 1 Review

The following problems are a mixture of addition, subtraction, multiplication, and division of whole numbers. Solve these problems. Take your time and work carefully.

1.

$$409 + 201$$

$$1024 - 14$$

$$138 + 328$$

$$560 - 245$$

2. $458 + 907 =$ _____ $265 - 95 =$ _____ $2209 + 1032 =$ _____

3. $237 - 125 =$ _____ $136 + 218 =$ _____ $4010 - 2400 =$ _____

4.

$$45 \times 18$$

$$308 \times 32$$

$$144 \times 12$$

$$312 \times 60$$

5. $15\overline{)155}$ $12\overline{)144}$ $32\overline{)280}$ $50\overline{)10{,}000}$

6. $88 \times 30 =$ _____ $175 \div 15 =$ _____ $25 \times 9 =$ _____

7. $145 \div 30 =$ _____ $51 \times 160 =$ _____ $256 \div 16 =$ _____

Estimate solutions for the following problems. Remember to round off one or more of the numbers as your first step.

8. 48 + 79 _____   201 – 9 _____   97 + 21 _____

9. 41 × 9 _____   198 ÷ 11 _____   25 × 99 _____

10. 15 + 109 _____   203 × 19 _____   97 ÷ 19 _____

 Use a calculator to do the following problems.

11. 252 ÷ 12 = _____   57 × 1,209 = _____   256 + 168 = _____

12. 565 – 143 = _____   355 × 16 = _____   2,561 – 186 = _____

Solve the following word problems. They may require addition, subtraction, multiplication, or division. Some may have extra information.

13. Mary Jo has 16 cassettes of video tape. Each cassette contains 20 minutes of footage. How many minutes of tape does Mary Jo have in all?

_____

14. Paul and Margaret drove 148 miles yesterday and 212 miles today. They plan to drive 200 miles tomorrow. How many miles did they drive the first two days?

_____

15. The 2 acres of Kay's Greenhouse and Gardens contain 25 rows of plants. If there are 25 plants in each row, how many plants does Kay have in all?

_____

16. Arianna weighed 8 pounds at birth. She gained 8 pounds during her first year and another 13 pounds her second year. How much did Arianna weigh at 2 years of age?

_____

17. Danny paid $47,590 for his house in 1993. He sold it in 1995 for $85,900. What was Danny's profit? (Hint: The profit is the difference between the price he paid when he bought the house and the price he got when he sold it.)

_____

Answers are on pages 136 to 137.

18. A chemist poured 114 ounces of liquid solution into glass beakers (see sample shown). How many beakers did she fill?

_____

3-ounce
Glass beaker

19. At Greg's Discount Auto Supply, there were 148 cans of oil on the shelf one Saturday. By the end of the following week, there were 51 cans remaining. How many cans of oil had been sold?

_____

20. There are 5,280 feet in a mile. How many feet are there in 8 miles?

_____

Unit 2 Review

The following problems are a mixture of addition, subtraction, multiplication, and division of decimals. Be sure to place the decimal point in the correct place in each answer.

1.
$$\begin{array}{r} 5.07 \\ +\ 2.8 \\ \hline \end{array}$$
$$\begin{array}{r} 10.5 \\ -\ 9.6 \\ \hline \end{array}$$
$$\begin{array}{r} 115.24 \\ +\ 100.09 \\ \hline \end{array}$$
$$\begin{array}{r} 20.9 \\ -\ 10 \\ \hline \end{array}$$

2. $2.009 + 5.45 =$ _____ $12.5 + 3.04 =$ _____

3. $.008 + 2.6 =$ _____ $5.075 + 10.59 =$ _____

4. $5.9 - .6 =$ _____ $4.003 - 2.1 =$ _____

5. $7 - 3.8 =$ _____ $200.9 - .175 =$ _____

6.
$$\begin{array}{r} 2.8 \\ \times\, 1.5 \\ \hline \end{array} \qquad \begin{array}{r} 40.1 \\ \times\quad .2 \\ \hline \end{array} \qquad \begin{array}{r} 3.01 \\ \times\quad 5 \\ \hline \end{array} \qquad \begin{array}{r} .09 \\ \times\, .32 \\ \hline \end{array}$$

7. $3.5\overline{)70.7}$ $.09\overline{)81}$ $.003\overline{).09}$

8. $9.9 \times 2 =$ _____ $5.08 \times 4.5 =$ _____ $16.6 \times .05 =$ _____

9. $99 \div .03 =$ _____ $.08 \div 4 =$ _____ $5.6 \div .8 =$ _____

Estimate solutions to the following problems. Round off
one or both numbers as a start.

10. $5.9 + 100.1 =$ _____ $2.01 - .9 =$ _____ $59.7 + 23.2 =$ _____

11. $6.6 \times 10.1 =$ _____ $2.99 \div .8 =$ _____ $4.009 \times 8.2 =$ _____

Use a calculator to do the following problems.

12. 56.19 + 90.4 = _____   12.8 − 4.55 = _____

13. 92.4 × 5.9 = _____   208.9 ÷ .85 = _____

Solve the following word problems. They may require addition, subtraction, multiplication, or division. Some may have unnecessary information.

14. The price of 2 pounds of ground beef is $4.40. How much does the meat cost per pound?

_____

15. Raoul mixed 8.5 ounces of soda water with 12.25 ounces of juice. He poured this mixture into a 24-ounce pitcher. How much liquid did he have in the pitcher?

_____

16. Payson worked 25.5 hours at a rate of $6.80 per hour. How much did he earn all together?

_____

17. There are 2.54 centimeters in 1 inch. How many centimeters are there in 10.5 inches?

_____

18. A dressmaker cut 12.7 inches from the fabric pictured below. How much was left of the fabric after she cut it?

_____

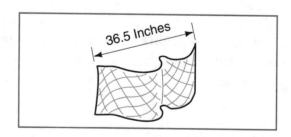

36.5 Inches

19. The Walk for Fitness Club covered the following miles:

June 28	4.7 miles
June 30	2.8 miles
July 1	6.5 miles
July 2	4.5 miles
July 3	3.9 miles

How many miles did the club members walk in all?

_____

20. The Manuel family's back lot is sketched below. How many square yards is the lot?

_____

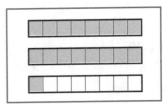

Answers are on pages 137 to 138.

Unit 3 Review

Write a fraction or mixed number for each figure below.

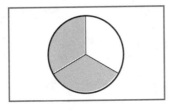 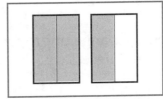 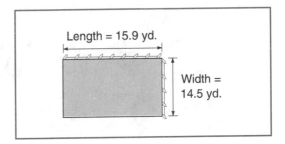

1. _____   _____   _____

Write a fraction for each situation described below. Reduce if possible.

2. Bannie worked 35 out of a possible 40 hours this week. What fraction of the week did Bannie work?

_____

3. Of every eighteen tires coming off the assembly line, two have something wrong with them. What fraction of the tires are not good?

_____

4. Five of the people pictured below are employed. What fraction is employed?

_____

5. A meter is equal to 100 centimeters. What fraction of a meter is 40 centimeters?

_____

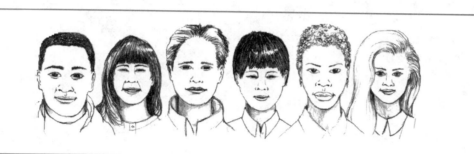

Find equivalent fractions for the following.

6. $\dfrac{4}{5} = \dfrac{}{20}$ $\dfrac{2}{3} = \dfrac{12}{}$ $\dfrac{9}{10} = \dfrac{}{90}$

Solve these fraction problems. Change improper fractions to mixed numbers and reduce if necessary.

7. $\dfrac{1}{3} + \dfrac{3}{4} = $ _____ $1\dfrac{7}{8} + 1\dfrac{3}{8} = $ _____ $2\dfrac{1}{4} + 4\dfrac{1}{2} = $ _____

8. $\dfrac{5}{8} - \dfrac{1}{4} = $ _____ $1\dfrac{1}{3} - \dfrac{2}{3} = $ _____ $6\dfrac{1}{8} - 2\dfrac{3}{4} = $ _____

9. $\dfrac{1}{2} \times \dfrac{5}{6} = $ _____ $8\dfrac{1}{2} \times 3\dfrac{1}{2} = $ _____ $4 \times 2\dfrac{3}{4} = $ _____

10. $\dfrac{3}{4} \div 2 = $ _____ $8\dfrac{1}{2} \div 2\dfrac{1}{2} = $ _____ $42 \div 5\dfrac{1}{4} = $ _____

Estimate an answer by rounding to the nearest whole number.

11. $3\frac{3}{4} + 5\frac{1}{8}$ _____ $9\frac{1}{10} - 3\frac{6}{7}$ _____ $2\frac{1}{5} + 3\frac{9}{10}$ _____

Solve these addition, subtraction, multiplication, and division word problems.

12. What is the perimeter (distance around) of the figure below?

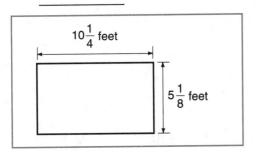

_____

10$\frac{1}{4}$ feet

5$\frac{1}{8}$ feet

13. What is the area of the figure above? (Hint: Multiply length by width.)

_____

14. Susan and Rob jog $4\frac{1}{2}$ miles each morning. If they jog 6 mornings each week, how many miles do they jog in a week?

_____

15. A truck can carry $3\frac{1}{8}$ tons of concrete. Harrison Excavating Company has $37\frac{1}{2}$ tons of concrete to haul. How many truckloads will it take to move it all?

_____

16. Annie Cohen Caterers had 15 pounds of fresh tomatoes at the start of the weekend. By Sunday night, there were $1\frac{1}{4}$ pounds left. How many pounds of tomatoes were used?

_____

17. Because of illness, Shatara worked only half of her scheduled hours this week. If she was scheduled for 37 hours, how many hours did she work?

_____

Answers are on pages 139 to 141.

Answers

Unit 1 Whole Numbers

LESSON 1   PLACE VALUE

Practice

1. *Answer given*	20	8,000
2. 8	0	3,000
3. 30	3	300
4. 90	100	50,000
5. 0	500	60,000

6. *Answer given*
7. 77
8. 8,305
9. 14,000
10. 44,610

LESSON 2   USING A CALCULATOR

Practice

1. 1,085   **2.** 1,933   **3.** 233   **4.** 1,440   **5.** 3,154

LESSON 3   ROUNDING AND ESTIMATING

Practice

1. *Answer given*	90	90	20
2. 10	50	10	70
3. *Answer given*	100	600	200
4. 100	500	200	100
5. *Answer given*	1,000	9,000	4,000
6. 1,000	6,000	4,000	8,000

LESSON 4   ADDING WHOLE NUMBERS

Practice

1.

$$\begin{array}{r} 4 \\ + 8 \\ \hline 12 \end{array} \qquad \begin{array}{r} 5 \\ + 6 \\ \hline 11 \end{array} \qquad \begin{array}{r} 3 \\ + 7 \\ \hline 10 \end{array} \qquad \begin{array}{r} 2 \\ + 6 \\ \hline 8 \end{array}$$

2.

$$\begin{array}{r} 9 \\ + 7 \\ \hline 16 \end{array} \qquad \begin{array}{r} 4 \\ + 5 \\ \hline 9 \end{array} \qquad \begin{array}{r} 1 \\ + 8 \\ \hline 9 \end{array} \qquad \begin{array}{r} 9 \\ + 5 \\ \hline 14 \end{array}$$

| **3.** | 34
+ 15
49 | 75
+ 12
87 | 90
+ 9
99 | 12
+ 47
59 | 17
+ 81
98 |

| **4.** | 83
+ 6
89 | 61
+ 18
79 | 44
+ 25
69 | 10
+ 88
98 | 11
+ 56
67 |

There may be more than one way to estimate answers 5 and 6.
Just be sure that your answers are close to the ones given here.

5. 140
36 rounds to 40
99 rounds to 100

40
14 rounds to 10
28 rounds to 30

50
37 rounds to 40
11 rounds to 10

6. 1,100
329 rounds to 300
790 rounds to 800

270
210 rounds to 200
66 rounds to 70

1,100
491 rounds to 500
550 rounds to 600

7. 2,561
2,586

8. 3,462
6,636

9. 64 ounces
32 + 32 = 64

10. 29 miles
18 + 11 = 29

11. 59 inches
51 + 8 = 59

12. $79
$27 + $40 + $12 = $79

LESSON 5   ADDING AND CARRYING

Practice

| **1.** | 60
+ 19
79 | $\overset{1}{2}8$
+ 12
40 | $\overset{1}{3}6$
+ 25
61 | 93
+ 21
114 | $\overset{1}{7}7$
+ 9
86 |

| **2.** | $\overset{1}{5}90$
+ 39
629 | $\overset{1}{2}48$
+ 102
350 | $\overset{11}{3},096$
+ 2,509
5,605 | $\overset{1}{2}34$
+ 219
453 | $\overset{1}{1}29$
+ 97
226 |

| **3.** | $\overset{1}{2}08$
+ 12
220 | 406
+ 321
727 | $\overset{11}{8}95$
+ 175
1,070 |

| **4.** | $\overset{1}{1}57$
+ 92
249 | $\overset{1}{9}71$
+ 333
1,304 | $\overset{1}{4}50$
+ 375
825 |

5. $40
$19 + $16 + $5 = $40

6. 832 tickets
634 + 198 = 832

7. 14,248 people
12,440 + 1,808 = 14,248

8. 36 miles
18 + 18 = 36

LESSON 6   SUBTRACTING WHOLE NUMBERS

Practice

1.

10	8	12	6
$-\ 2$	-7	$-\ 4$	-3
8	1	8	3

2.

9	14	8	9
-7	$-\ 5$	-3	-5
2	9	5	4

3.

38	75	99	42	87
-15	-12	$-\ 9$	-41	-21
23	63	90	1	66

4.

34	61	44	75	11
-22	$-\ 1$	-32	-22	$-\ 9$
12	60	12	53	2

5. 25
87 − 50 = 37
37 − 12 = 25

211
368 − 124 = 244
244 − 33 = 211

6. 120 pages
244 − 124 = 120

7. 112 fish
145 − 33 = 112

8. 224 days
365 − 141 = 224

9. 130 miles
550 − 420 = 130

10. 113 centimeters
145 − 32 = 113

LESSON 7   SUBTRACTING AND BORROWING

Practice

1.

$5^4 8$	$1\ 7^6 5$	$9^8 3$	$4^3 2$	$8^7 0$
$-1\ 9$	$-1\ 4\ 6$	$-\ 9$	$-2\ 8$	$-2\ 1$
$3\ 9$	$2\ 9$	$8\ 4$	$1\ 4$	$5\ 9$

2.

$1\ 3^2 4$	$6^5 0$	$2\ 6^5 4$	$7^6 5$	$4^3 1$
$-\ 2\ 6$	$-\ 1$	$-1\ 3\ 9$	$-2\ 9$	$-1\ 8$
$1\ 0\ 8$	$5\ 9$	$1\ 2\ 5$	$4\ 6$	$2\ 3$

3.
$$\begin{array}{r} 2,9\overset{8}{\cancel{9}}\overset{10}{\cancel{1}}2 \\ -1,573 \\ \hline 1,339 \end{array}$$
$$\begin{array}{r} \overset{1}{\cancel{2}}\overset{9}{\cancel{0}}4 \\ -139 \\ \hline 65 \end{array}$$
$$\begin{array}{r} \overset{3}{\cancel{4}}\overset{9}{\cancel{0}}5 \\ -316 \\ \hline 89 \end{array}$$
$$\begin{array}{r} 1,\overset{9}{\cancel{0}}39 \\ -981 \\ \hline 58 \end{array}$$
$$\begin{array}{r} 1\overset{1}{\cancel{2}}3 \\ -55 \\ \hline 68 \end{array}$$

There may be more than one way to estimate answers 4 and 5.
Just be sure that your answers are close to the ones given here.

4. 30
    36 rounds to 40
    14 rounds to 10

0
    14 rounds to 10
    9 rounds to 10

30
    37 rounds to 40
    11 rounds to 10

5. 270
    349 rounds to 300
    29 rounds to 30

190
    210 rounds to 200
    9 rounds to 10

450
    499 rounds to 500
    51 rounds to 50

6. 1,655   12,287

7. 79   36

8. $480
    $1,040 − $560 = $480

9. 27 ounces
    64 − 37 = 27

10. 27 pounds
    50 − 23 = 27

11. 105 more people
    1,050 − 945 = 105

LESSON 8   MULTIPLYING WHOLE NUMBERS

Practice

1.
$$\begin{array}{r} 71 \\ \times\ 8 \\ \hline 568 \end{array}$$
$$\begin{array}{r} 111 \\ \times\ 5 \\ \hline 555 \end{array}$$
$$\begin{array}{r} 34 \\ \times\ 2 \\ \hline 68 \end{array}$$
$$\begin{array}{r} 601 \\ \times\ 9 \\ \hline 5,409 \end{array}$$
$$\begin{array}{r} 103 \\ \times\ 3 \\ \hline 309 \end{array}$$

2.
$$\begin{array}{r} 21 \\ \times\ 7 \\ \hline 147 \end{array}$$
$$\begin{array}{r} 143 \\ \times\ 2 \\ \hline 286 \end{array}$$
$$\begin{array}{r} 642 \\ \times\ 2 \\ \hline 1,284 \end{array}$$
$$\begin{array}{r} 420 \\ \times\ 4 \\ \hline 1,680 \end{array}$$
$$\begin{array}{r} 101 \\ \times\ 8 \\ \hline 808 \end{array}$$

3.
$$\begin{array}{r} 11 \\ \times\ 78 \\ \hline 88 \\ 77\ \\ \hline 858 \end{array}$$
$$\begin{array}{r} 221 \\ \times\ 33 \\ \hline 663 \\ 6\,63\ \\ \hline 7,293 \end{array}$$
$$\begin{array}{r} 34 \\ \times\ 22 \\ \hline 68 \\ 68\ \\ \hline 748 \end{array}$$
$$\begin{array}{r} 301 \\ \times\ 19 \\ \hline 2,709 \\ 3\,01\ \\ \hline 5,719 \end{array}$$
$$\begin{array}{r} 143 \\ \times\ 23 \\ \hline 429 \\ 2\,86\ \\ \hline 3,289 \end{array}$$

4.
$$\begin{array}{r} 12 \\ \times\ 12 \\ \hline 24 \\ 12\ \\ \hline 144 \end{array}$$
$$\begin{array}{r} 211 \\ \times\ 11 \\ \hline 211 \\ 2\,11\ \\ \hline 2,321 \end{array}$$
$$\begin{array}{r} 44 \\ \times\ 22 \\ \hline 88 \\ 88\ \\ \hline 968 \end{array}$$
$$\begin{array}{r} 312 \\ \times\ 13 \\ \hline 936 \\ 3\,12\ \\ \hline 4,056 \end{array}$$
$$\begin{array}{r} 103 \\ \times\ 23 \\ \hline 309 \\ 2\,06\ \\ \hline 2,369 \end{array}$$

5. 396 miles
$$\begin{array}{r} 33 \\ \times\ 12 \\ \hline 66 \\ 33\ \\ \hline 396 \end{array}$$

6. 300 minutes
    50 × 6 = 300

7. 288 cans
$$\begin{array}{r} 24 \\ \times\ 12 \\ \hline 48 \\ 24\ \\ \hline 288 \end{array}$$

8. 60 inches
    5 × 12 = 60

Practice

1.
$$\overset{1}{2}3 \\ \times\ 4 \\ \overline{92}$$

$$\overset{4}{4}5 \\ \times\ 9 \\ \overline{405}$$

$$9\overset{1}{0}3 \\ \times\ \ 6 \\ \overline{5,418}$$

$$2\overset{2}{1}7 \\ \times\ \ 3 \\ \overline{651}$$

2.
$$1\overset{3}{0}7 \\ \times\ \ 5 \\ \overline{535}$$

$$3\overset{5}{7} \\ \times\ 8 \\ \overline{296}$$

$$50 \\ \times\ 7 \\ \overline{350}$$

3.
$$125 \\ \times\ 15 \\ \overline{625} \\ \underline{1\ 25} \\ 1,875$$

$$500 \\ \times\ \ 34 \\ \overline{2,000} \\ \underline{15\ 00} \\ 17,000$$

$$309 \\ \times\ \ 25 \\ \overline{1,545} \\ \underline{6\ 18} \\ 7,725$$

$$112 \\ \times\ 10 \\ \overline{000} \\ \underline{1\ 12} \\ 1,120$$

$$240 \\ \times\ 22 \\ \overline{480} \\ \underline{4\ 80} \\ 5,280$$

4.
$$225 \\ \times\ 30 \\ \overline{000} \\ \underline{6\ 75} \\ 6,750$$

$$210 \\ \times\ 15 \\ \overline{1050} \\ \underline{2\ 10} \\ 3,150$$

$$300 \\ \times\ 42 \\ \overline{600} \\ \underline{12\ 00} \\ 12,600$$

$$512 \\ \times\ 10 \\ \overline{000} \\ \underline{5\ 12} \\ 5,120$$

$$245 \\ \times\ 20 \\ \overline{000} \\ \underline{4\ 90} \\ 4,900$$

5.
$$150 \\ \times\ 35 \\ \overline{750} \\ \underline{4\ 50} \\ 5,250$$

$$278 \\ \times\ 12 \\ \overline{556} \\ \underline{2\ 78} \\ 3,336$$

$$256 \\ \times\ 18 \\ \overline{2048} \\ \underline{2\ 56} \\ 4,608$$

There are many possible ways to estimate the answers to 6.
Just be sure that your answers are close to the ones given here.

6. 4,000
405 rounds to 400
11 rounds to 10

$$400 \\ \times\ 10 \\ \overline{000} \\ \underline{4\ 00} \\ 4,000$$

4,000
198 rounds to 200
19 rounds to 20

$$200 \\ \times\ 20 \\ \overline{000} \\ \underline{4\ 00} \\ 4,000$$

4,000
35 rounds to 40
121 rounds to 100

$$100 \\ \times\ 40 \\ \overline{000} \\ \underline{4\ 00} \\ 4,000$$

7. 5,250   6,480   2,900

8. 420 salads
$$35 \\ \times 12 \\ \overline{70} \\ \underline{35} \\ 420$$

9. 1,715 passengers
$245 \times 7 = 1,715$

10. $2,100
$$175 \\ \times\ 12 \\ \overline{350} \\ \underline{1\ 75} \\ 2,100$$

Answers

11. 770 milliliters

$$
\begin{array}{r}
55 \\
\times\,14 \\
\hline
220 \\
55 \\
\hline
770
\end{array}
$$

12. 1,350 square yards

$$
\begin{array}{r}
90 \\
\times\ 15 \\
\hline
450 \\
90 \\
\hline
1{,}350
\end{array}
$$

LESSON 10   DIVIDING WHOLE NUMBERS

Practice

1. 5 7×5

2. 8 3×8

3. 24   12   102   54

$$
\begin{array}{r}
24 \\
5\overline{)120} \\
\underline{10} \\
20 \\
\underline{20} \\
0
\end{array}
\qquad
\begin{array}{r}
12 \\
3\overline{)36} \\
\underline{3} \\
06 \\
\underline{6} \\
0
\end{array}
\qquad
\begin{array}{r}
102 \\
8\overline{)816} \\
\underline{8} \\
016 \\
\underline{16} \\
0
\end{array}
\qquad
\begin{array}{r}
54 \\
4\overline{)216} \\
\underline{20} \\
16 \\
\underline{16} \\
0
\end{array}
$$

4. 31   56   21

$$
\begin{array}{r}
31 \\
5\overline{)155} \\
\underline{15} \\
05 \\
\underline{5} \\
0
\end{array}
\qquad
\begin{array}{r}
56 \\
4\overline{)224} \\
\underline{20} \\
24 \\
\underline{24} \\
0
\end{array}
\qquad
\begin{array}{r}
21 \\
8\overline{)168} \\
\underline{16} \\
08 \\
\underline{8} \\
0
\end{array}
$$

5. 34r3   17   56

$$
\begin{array}{r}
34 \\
8\overline{)275} \\
\underline{24} \\
35 \\
\underline{32} \\
3
\end{array}
\qquad
\begin{array}{r}
17 \\
9\overline{)153} \\
\underline{9} \\
63 \\
\underline{63} \\
0
\end{array}
\qquad
\begin{array}{r}
56 \\
8\overline{)448} \\
\underline{40} \\
48 \\
\underline{48} \\
0
\end{array}
$$

6. Three slices

$12 \div 4 = 3$

7. 322 miles

$966 \div 3 = 322$

8. 42 feet

$252 \div 6 = 42$

Practice

1. 22

$$
\begin{array}{r}
22 \\
20\overline{)440} \\
\underline{40} \\
40 \\
\underline{40} \\
0
\end{array}
$$

25

$$
\begin{array}{r}
25 \\
15\overline{)375} \\
\underline{30} \\
75 \\
\underline{75} \\
0
\end{array}
$$

19

$$
\begin{array}{r}
19 \\
38\overline{)722} \\
\underline{38} \\
342 \\
\underline{342} \\
0
\end{array}
$$

12

$$
\begin{array}{r}
12 \\
12\overline{)144} \\
\underline{12} \\
24 \\
\underline{24} \\
0
\end{array}
$$

2. 20

$$
\begin{array}{r}
20 \\
15\overline{)300} \\
\underline{30} \\
00 \\
\underline{0} \\
0
\end{array}
$$

21

$$
\begin{array}{r}
21 \\
13\overline{)273} \\
\underline{26} \\
13 \\
\underline{13} \\
0
\end{array}
$$

20

$$
\begin{array}{r}
20 \\
17\overline{)340} \\
\underline{34} \\
00 \\
\underline{0} \\
0
\end{array}
$$

3. 15r5

$$
\begin{array}{r}
15 \\
18\overline{)275} \\
\underline{18} \\
95 \\
\underline{90} \\
5
\end{array}
$$

8r4

$$
\begin{array}{r}
8 \\
17\overline{)140} \\
\underline{136} \\
4
\end{array}
$$

5r48

$$
\begin{array}{r}
5 \\
80\overline{)448} \\
\underline{400} \\
48
\end{array}
$$

4. 4r21

$$
\begin{array}{r}
4 \\
33\overline{)153} \\
\underline{132} \\
21
\end{array}
$$

18

$$
\begin{array}{r}
18 \\
20\overline{)360} \\
\underline{20} \\
160 \\
\underline{160} \\
0
\end{array}
$$

32r4

$$
\begin{array}{r}
32 \\
16\overline{)516} \\
\underline{48} \\
36 \\
\underline{32} \\
4
\end{array}
$$

5. 18 pints
$288 \div 16 = 18$

6. 48 crates
$1{,}152 \div 24 = 48$

7. 12 hours
$660 \div 55 = 12$

8. $3
$42 \div 14 = \$3$

Unit 2 *Decimals*

LESSON 12   WHAT ARE DECIMALS?

Practice

1. .75   **2.** .5   **3.** 1.1   **4.** .9

5. 5; One thousand two hundred ninety-eight and one hundred forty-five thousandths

6. 6; Four hundred twenty-one and sixteen hundredths

7. tenths; Eight thousand nine hundred sixty-one and forty-five thousandths

8. ones; Nine hundred eighty-seven and twelve hundredths

LESSON 13   ZEROS IN DECIMALS

Practice

1. 35.075   **2.** .350   **3.** .98   **4.** 7.02

5. .88   **6.** .9

LESSON 14   ROUNDING AND ESTIMATING

Practice

1. 5
4.98 rounds to 5

125
125.03 rounds to 125

36
35.5 rounds to 36

2. 10
9.91 rounds to 10

80
80.4 rounds to 80

13
12.7 rounds to 13

3. 10.3
10.25 rounds to 10.3

3.1
3.08 rounds to 3.1

24.0
24.02 rounds to 24.0

4. 2.1
2.11 rounds to 2.1

6.5
6.48 rounds to 6.5

1.6
1.55 rounds to 1.6

There are many possible ways to estimate answers 5 to 8. Just be sure that your answers are close to the ones given here.

5. $12
$3 + $5 = $8
$2.89 rounds to $3
$5.12 rounds to $5
$20 − $8 = $12

6. $300
$700 − $400 = $300
$678.98 rounds to $700
$404.50 rounds to $400

7. No
$17 + $18 = $35
$17.03 rounds to $17
$17.55 rounds to $18

8. $200
8 × $25 = $200
$24.99 rounds to $25

9. $1.40
$1.399 rounds to $1.40

10. 12.8 pounds
12.78 rounds to 12.8 pounds

11. 13 pounds
12.78 rounds to 13 pounds

12. 24,875 miles
24,875.4 rounds to 24,875 miles

Practice

1.
$$\begin{array}{r} \overset{1}{5.5} \\ +3.8 \\ \hline 9.3 \end{array} \qquad \begin{array}{r} 9.9 \\ -1.5 \\ \hline 8.4 \end{array} \qquad \begin{array}{r} \overset{1}{125.75} \\ +\ 10.50 \\ \hline 136.25 \end{array} \qquad \begin{array}{r} 78.2 \\ -\ 3.0 \\ \hline 75.2 \end{array}$$

2.
$$\begin{array}{r} \overset{1}{8.5} \\ +\ 2.5 \\ \hline 11.0 \end{array} \qquad \begin{array}{r} \overset{1}{2}.\overset{1}{1} \\ -1.9 \\ \hline .2 \end{array} \qquad \begin{array}{r} \overset{1}{25.02} \\ +\ 8.90 \\ \hline 33.92 \end{array} \qquad \begin{array}{r} 1\overset{7}{8}.\overset{1}{2} \\ -\ 5.9 \\ \hline 12.3 \end{array}$$

3.
$$\begin{array}{r} \overset{1}{3.50} \\ +\ 9.89 \\ \hline 13.39 \end{array} \qquad \begin{array}{r} \overset{1}{1.80} \\ +5.75 \\ \hline 7.55 \end{array} \qquad \begin{array}{r} 3.03 \\ +1.25 \\ \hline 4.28 \end{array}$$

4.
$$\begin{array}{r} 10.\overset{4}{5}\overset{1}{0} \\ -\ 8.25 \\ \hline 2.25 \end{array} \qquad \begin{array}{r} 9.\overset{3}{4}\overset{1}{0} \\ -2.05 \\ \hline 7.35 \end{array} \qquad \begin{array}{r} 5.25 \\ -4.10 \\ \hline 1.15 \end{array}$$

5. .08 increase
.355 − .275 = .08

6. 6.8 miles
1.5 + 1.9 + 3.4 = 6.8

7. $6.03
$20 + $10 = $30
$30 − $23.97 = $6.03

8. 3.8 degrees
102.4 − 98.6 = 3.8

9. 22.3 acres
13.8 + 8.5 = 22.3

10. 9.25 seconds
5.75 + 3.5 = 9.25

Practice

1. 320.8
$$\begin{array}{r} 100.25 \\ \times\quad 3.20 \\ \hline 00000 \\ 20050\ \ \\ 30075\quad \\ \hline 320.8000 \end{array} \qquad \begin{array}{r} 1.518 \\ 75.90 \\ \times\quad .02 \\ \hline 15180 \\ 0000\ \ \\ \hline 1.5180 \end{array} \qquad \begin{array}{r} 10.89 \\ 9.9 \\ \times\ 1.1 \\ \hline 99 \\ 99\ \ \\ \hline 10.89 \end{array} \qquad \begin{array}{r} .008 \\ .04 \\ \times\ .20 \\ \hline 00 \\ 08\ \ \\ \hline .0080 \end{array}$$

2. 10.45
$$\begin{array}{r} 5.5 \\ \times\ 1.9 \\ \hline 495 \\ 55\ \ \\ \hline 10.45 \end{array} \qquad \begin{array}{r} 45.45 \\ 90.9 \\ \times\quad .5 \\ \hline 45.45 \end{array} \qquad \begin{array}{r} 72.16 \\ 16.4 \\ \times\ 4.4 \\ \hline 656 \\ 656\ \ \\ \hline 72.16 \end{array}$$

There are many possible ways to estimate answers 3 to 4.
Just be sure that your answers are close to the ones given here.

3. 100
    99.9 rounds to 100
    1.1 rounds to 1

    78
    25.5 rounds to 26
    3.01 rounds to 3

    50
    4.8 rounds to 5
    9.7 rounds to 10

4. 1,800
    9.2 rounds to 9
    200.1 rounds to 200

    210
    29.5 rounds to 30
    6.99 rounds to 7

    15
    3.25 rounds to 3
    4.8 rounds to 5

5. 14.8225   128.25   46.875

6. 17.2995   12.54   .0045

7. $1.47
    $3 \times \$.49 = \$1.47$

8. 126 tons
    $10.5 \times 12 = 126$

9. $4.50
    $\$1.80 \times 2.5 = \$4.50$

10. 61.25 kilograms
    $35 \times 1.75 = 61.25$

11. 69.375 square feet
    $7.5 \times 9.25 = 69.375$

12. $8.25
    $\$5.50 \times 1.5 = \$8.25$

LESSON 17   DIVIDING DECIMALS BY A WHOLE NUMBER

Practice

1. 2.501

$$
\begin{array}{r}
2.501 \\
5\overline{)12.505} \\
\underline{10} \\
2\,5 \\
\underline{2\,5} \\
005 \\
\underline{5} \\
0
\end{array}
$$

.304

$$
\begin{array}{r}
.304 \\
10\overline{)3.040} \\
\underline{3\,0} \\
040 \\
\underline{40} \\
0
\end{array}
$$

.002

$$
\begin{array}{r}
.002 \\
4\overline{)0.008} \\
\underline{8} \\
0
\end{array}
$$

2. .704

$$
\begin{array}{r}
.704 \\
3\overline{)2.112} \\
\underline{2\,1} \\
012 \\
\underline{12} \\
0
\end{array}
$$

.103

$$
\begin{array}{r}
.103 \\
12\overline{)1.236} \\
\underline{1\,2} \\
036 \\
\underline{36} \\
0
\end{array}
$$

.96

$$
\begin{array}{r}
.96 \\
6\overline{)5.76} \\
\underline{5\,4} \\
36 \\
\underline{36} \\
0
\end{array}
$$

3. 30.101

$$
\begin{array}{r}
30.101 \\
10\overline{)301.010} \\
\underline{30} \\
01\,0 \\
\underline{1\,0} \\
010 \\
\underline{10} \\
0
\end{array}
$$

.093

$$
\begin{array}{r}
.093 \\
3\overline{).279} \\
\underline{27} \\
09 \\
\underline{9} \\
0
\end{array}
$$

.36

$$
\begin{array}{r}
.36 \\
5\overline{)1.80} \\
\underline{1\,5} \\
30 \\
\underline{30} \\
0
\end{array}
$$

There are many possible ways to estimate the answers in 4.
Just be sure that your answers are close to the ones given here.

4. 17   4   2
    50.9 rounds to 51   8.01 rounds to 8   9.98 rounds to 10
    3.1 rounds to 3   1.9 rounds to 2   4.8 rounds to 5

5. Four lengths   **6.** $12.90
    102 ÷ 25.5 = 4 $11.40 + $14.40 = $25.80
                              $25.80 ÷ 2 = $12.90

7. $9.02   **8.** 8.4 inches
    $36.08 ÷ 4 = $9.02   25.2 ÷ 3 = 8.4

9. 5.09 grams   **10.** $14.30
    25.45 ÷ 5 = 5.09 $14.50 + $18.00 + $10.40 = $42.90
                              $42.90 ÷ 3 = $14.30

LESSON 18   DIVIDING A DECIMAL BY A DECIMAL

Practice

1. 23   20.3   2,000

$$.25\overline{)5.75\,0}$$ gives 23.0
$$1.4\overline{)28.4\,2}$$ gives 20.3
$$.035\overline{)70.000}$$ gives 2 000

```
        23.0              2 0.3              2 000
.25 )5.75 0        1.4 )28.4 2        .035 )70.000
     5 0               28                  70
     75               0 42               0 000
     75                4 2
      0 0                0
```

2. 45   .3   30.3

```
        4 5                 .3                3 0.3
.5 )22.5           .03 ).00 9         2.5 )75.7 5
   20                  9                  75
   2 5                 0                 0 7 5
   2 5                                    7 5
    0                                      0
```

3. 1.7   .4   6

4. 10,000   1.025   .5

5. $7.50   **6.** 14 bunches
    $266.25 ÷ 35.5 = $7.50 $12.46 ÷ $.89 = 14

7. 28.5 miles per gallon   **8.** 40 apples
    307.8 ÷ 10.8 = 28.5 $6 ÷ $.15 = 40

Unit 3 Fractions

LESSON 19   PARTS OF A WHOLE

Practice

1. $\dfrac{3}{4}$ $\qquad$ $\dfrac{5}{6}$ $\qquad$ $\dfrac{1}{10}$

2. $\dfrac{2}{10}$ $\qquad$ $\dfrac{1}{2}$ $\qquad$ $\dfrac{2}{5}$

3. $\dfrac{19}{36}$ $\qquad$ **4.** $\dfrac{4}{5}$ $\qquad$ **5.** $\dfrac{43}{100}$ $\qquad$ **6.** $\dfrac{3}{10}$

7. $\dfrac{31}{105}$ $\qquad$ **8.** $\dfrac{13}{60}$

LESSON 20   EQUIVALENT FRACTIONS

Practice

1. $\dfrac{3}{5} = \dfrac{9}{15}$ $\qquad$ $\dfrac{1}{8} = \dfrac{3}{24}$ $\qquad$ $\dfrac{12}{15} = \dfrac{4}{5}$

   $\dfrac{3}{5} \times \dfrac{3}{3} = \dfrac{9}{15}$ $\qquad$ $\dfrac{1}{8} \times \dfrac{3}{3} = \dfrac{3}{24}$ $\qquad$ $\dfrac{12}{15} \div \dfrac{3}{3} = \dfrac{4}{5}$

2. $\dfrac{6}{8} = \dfrac{3}{4}$ $\qquad$ $\dfrac{2}{3} = \dfrac{6}{9}$ $\qquad$ $\dfrac{24}{48} = \dfrac{6}{12}$

   $\dfrac{6}{8} \div \dfrac{2}{2} = \dfrac{3}{4}$ $\qquad$ $\dfrac{2}{3} \times \dfrac{3}{3} = \dfrac{6}{9}$ $\qquad$ $\dfrac{24}{48} \div \dfrac{4}{4} = \dfrac{6}{12}$

3. $\dfrac{1}{6} = \dfrac{5}{30}$ $\qquad$ $\dfrac{4}{7} = \dfrac{12}{21}$ $\qquad$ $\dfrac{1}{8} = \dfrac{4}{32}$

   $\dfrac{1}{6} \times \dfrac{5}{5} = \dfrac{5}{30}$ $\qquad$ $\dfrac{4}{7} \times \dfrac{3}{3} = \dfrac{12}{21}$ $\qquad$ $\dfrac{1}{8} \times \dfrac{4}{4} = \dfrac{4}{32}$

4. $\dfrac{9}{18} = \dfrac{1}{2}$ $\qquad$ $\dfrac{3}{4} = \dfrac{30}{40}$ $\qquad$ $\dfrac{5}{9} = \dfrac{15}{27}$

   $\dfrac{9}{18} \div \dfrac{9}{9} = \dfrac{1}{2}$ $\qquad$ $\dfrac{3}{4} \div \dfrac{10}{10} = \dfrac{30}{40}$ $\qquad$ $\dfrac{5}{9} \times \dfrac{3}{3} = \dfrac{15}{27}$

5. $\dfrac{6}{8}$ of a cup $\qquad\qquad$ **6.** $\dfrac{4}{5}$ of a mile

   $\dfrac{3}{4} \times \dfrac{2}{2} = \dfrac{6}{8}$ $\qquad\qquad$ $\dfrac{8}{10} \div \dfrac{2}{2} = \dfrac{4}{5}$

7. $\dfrac{3}{4}$ of a pound $\qquad\qquad$ **8.** $\dfrac{8}{12}$ of a foot

   $\dfrac{12}{16} \div \dfrac{4}{4} = \dfrac{3}{4}$ $\qquad\qquad$ $\dfrac{2}{3} \times \dfrac{4}{4} = \dfrac{8}{12}$

125

LESSON 21 ROUNDING AND ESTIMATING

Practice

1. $\dfrac{7}{14}$ **2.** $\dfrac{1}{8}, \dfrac{2}{21}$ **3.** $\dfrac{5}{6}, \dfrac{9}{11}, \dfrac{12}{13}$ **4.** $\dfrac{10}{21}$

The answers in 5 to 7 are examples. Your answers may be different.

5. $\dfrac{8}{9}$ $\dfrac{10}{11}$ $\dfrac{9}{10}$ $\dfrac{6}{7}$ $\dfrac{4}{5}$ $\dfrac{9}{10}$ $\dfrac{3}{4}$

6. $\dfrac{3}{6}$ $\dfrac{5}{10}$ $\dfrac{5}{10}$ $\dfrac{2}{4}$ $\dfrac{4}{8}$ $\dfrac{9}{18}$ $\dfrac{4}{9}$

7. $\dfrac{3}{11}$ $\dfrac{9}{23}$ $\dfrac{1}{100}$ $\dfrac{2}{13}$ $\dfrac{1}{85}$ $\dfrac{6}{23}$ $\dfrac{1}{13}$

8. She is almost finished.
9. She has read a small fraction of the book.
10. He has traveled about one half of the trip.

LESSON 22 COMMON DENOMINATORS

Practice

1. $5 \times 8 = 40$

$\dfrac{4}{5} \times \dfrac{8}{8} = \dfrac{32}{40}$

$\dfrac{1}{8} \times \dfrac{5}{5} = \dfrac{5}{40}$

$10 \div 5 = 2$

$\dfrac{1}{10} \times \dfrac{1}{1} = \dfrac{1}{10}$

$\dfrac{3}{5} \times \dfrac{2}{2} = \dfrac{6}{10}$

$7 \times 4 = 28$

$\dfrac{3}{7} \times \dfrac{4}{4} = \dfrac{12}{28}$

$\dfrac{1}{4} \times \dfrac{7}{7} = \dfrac{7}{28}$

2. 6 & 8 are multiples of 24

$\dfrac{1}{6} \times \dfrac{4}{4} = \dfrac{4}{24}$

$\dfrac{1}{8} \times \dfrac{3}{3} = \dfrac{3}{24}$

$3 \times 4 = 12$

$\dfrac{2}{3} \times \dfrac{4}{4} = \dfrac{8}{12}$

$\dfrac{3}{4} \times \dfrac{3}{3} = \dfrac{9}{12}$

$16 \div 8 = 2$

$\dfrac{5}{8} \times \dfrac{2}{2} = \dfrac{10}{16}$

$\dfrac{1}{16} \times \dfrac{1}{1} = \dfrac{1}{16}$

LESSON 23 ADDING FRACTIONS

Practice

1. $\dfrac{1}{3} + \dfrac{2}{3} = \dfrac{3}{3} = 1$ $\dfrac{3}{8} + \dfrac{1}{8} = \dfrac{4}{8} = \dfrac{1}{2}$ $\dfrac{1}{5} + \dfrac{3}{5} = \dfrac{4}{5}$

2. $1\dfrac{1}{6}$

$\dfrac{1}{2} = \dfrac{3}{6}$

$\dfrac{2}{3} = \dfrac{4}{6}$

$\dfrac{3}{6} + \dfrac{4}{6} = 1\dfrac{1}{6}$

$\dfrac{7}{16}$

$\dfrac{3}{8} = \dfrac{6}{16}$

$\dfrac{1}{16} = \dfrac{1}{16}$

$\dfrac{6}{16} + \dfrac{1}{16} = \dfrac{7}{16}$

$1\dfrac{3}{28}$

$\dfrac{6}{7} = \dfrac{24}{28}$

$\dfrac{1}{4} = \dfrac{7}{28}$

$\dfrac{24}{28} + \dfrac{7}{28} = \dfrac{31}{28}$

$\dfrac{31}{28} = 1\dfrac{3}{28}$

3. $1\frac{1}{12}$

$\frac{3}{4} = \frac{9}{12}$

$\frac{1}{3} = \frac{4}{12}$

$\frac{9}{12} + \frac{4}{12} = \frac{13}{12}$

$\frac{13}{12} = 1\frac{1}{12}$

$\frac{9}{10}$

$\frac{1}{10} = \frac{1}{10}$

$\frac{4}{5} = \frac{8}{10}$

$\frac{1}{10} + \frac{8}{10} = \frac{9}{10}$

$\frac{4}{9}$

$\frac{1}{9} = \frac{1}{9}$

$\frac{1}{3} = \frac{3}{9}$

$\frac{1}{9} + \frac{3}{9} = \frac{4}{9}$

4. $\frac{7}{12}$ yard

$\frac{1}{3} = \frac{4}{12}$

$\frac{1}{4} = \frac{3}{12}$

$\frac{4}{12} + \frac{3}{12} = \frac{7}{12}$

5. $1\frac{3}{8}$ pounds

$\frac{3}{4} = \frac{6}{8}$

$\frac{5}{8} = \frac{5}{8}$

$\frac{6}{8} + \frac{5}{8} = \frac{11}{8}$

$\frac{11}{8} = 1\frac{3}{8}$

6. $1\frac{1}{2}$ miles

$\frac{3}{5} = \frac{6}{10}$

$\frac{9}{10} = \frac{9}{10}$

$\frac{6}{10} + \frac{9}{10} = \frac{15}{10}$

$\frac{15}{10} = 1\frac{5}{10} = 1\frac{1}{2}$

7. $\frac{5}{16}$ centimeter

$\frac{3}{16} = \frac{3}{16}$

$\frac{1}{8} = \frac{2}{16}$

$\frac{3}{16} + \frac{2}{16} = \frac{5}{16}$

8. $\frac{1}{2}$ of the folders

$\frac{1}{4} + \frac{1}{4} = \frac{2}{4}$

$\frac{2}{4} = \frac{1}{2}$

LESSON 24   SUBTRACTING FRACTIONS

Practice

1. $\frac{4}{9}$

$\frac{7}{9} - \frac{3}{9} = \frac{4}{9}$

$\frac{5}{6}$

$\frac{11}{12} - \frac{1}{12} = \frac{10}{12}$

$\frac{10}{12} = \frac{5}{6}$

$\frac{2}{3}$

$\frac{5}{6} - \frac{1}{6} = \frac{4}{6}$

$\frac{4}{6} = \frac{2}{3}$

2. $\dfrac{5}{8}$ $\qquad$ $\dfrac{1}{2}$ $\qquad$ $\dfrac{1}{4}$

$\dfrac{7}{8} = \dfrac{7}{8}$ $\qquad$ $\dfrac{2}{3} = \dfrac{4}{6}$ $\qquad$ $\dfrac{7}{12} = \dfrac{7}{12}$

$\dfrac{1}{4} = \dfrac{2}{8}$ $\qquad$ $\dfrac{1}{6} = \dfrac{1}{6}$ $\qquad$ $\dfrac{1}{3} = \dfrac{4}{12}$

$\dfrac{7}{8} - \dfrac{2}{8} = \dfrac{5}{8}$ $\qquad$ $\dfrac{4}{6} - \dfrac{1}{6} = \dfrac{3}{6}$ $\qquad$ $\dfrac{7}{12} - \dfrac{4}{12} = \dfrac{3}{12}$

$\qquad\qquad\qquad$ $\dfrac{3}{6} = \dfrac{1}{2}$ $\qquad$ $\dfrac{3}{12} = \dfrac{1}{4}$

3. $\dfrac{1}{2}$ $\qquad$ $\dfrac{1}{2}$ $\qquad$ $\dfrac{1}{9}$

$\dfrac{4}{5} = \dfrac{8}{10}$ $\qquad$ $\dfrac{13}{14} = \dfrac{13}{14}$ $\qquad$ $\dfrac{1}{3} = \dfrac{3}{9}$

$\dfrac{3}{10} = \dfrac{3}{10}$ $\qquad$ $\dfrac{3}{7} = \dfrac{6}{14}$ $\qquad$ $\dfrac{2}{9} = \dfrac{2}{9}$

$\dfrac{8}{10} - \dfrac{3}{10} = \dfrac{5}{10}$ $\qquad$ $\dfrac{13}{14} - \dfrac{6}{14} = \dfrac{7}{14}$ $\qquad$ $\dfrac{3}{9} - \dfrac{2}{9} = \dfrac{1}{9}$

$\dfrac{5}{10} = \dfrac{1}{2}$ $\qquad$ $\dfrac{7}{14} = \dfrac{1}{2}$

4. $\dfrac{1}{5}$ $\qquad$ $\dfrac{5}{6}$ $\qquad$ $\dfrac{1}{4}$

$1 = \dfrac{5}{5}$ $\qquad$ $1 = \dfrac{6}{6}$ $\qquad$ $1 = \dfrac{4}{4}$

$\dfrac{4}{5} = \dfrac{4}{5}$ $\qquad$ $\dfrac{1}{6} = \dfrac{1}{6}$ $\qquad$ $\dfrac{3}{4} = \dfrac{3}{4}$

$\dfrac{5}{5} - \dfrac{4}{5} = \dfrac{1}{5}$ $\qquad$ $\dfrac{6}{6} - \dfrac{1}{6} = \dfrac{5}{6}$ $\qquad$ $\dfrac{4}{4} - \dfrac{3}{4} = \dfrac{1}{4}$

5. $\dfrac{5}{12}$ cup

$\dfrac{3}{4} = \dfrac{9}{12}$

$\dfrac{1}{3} = \dfrac{4}{12}$

$\dfrac{9}{12} - \dfrac{4}{12} = \dfrac{5}{12}$

6. $\dfrac{2}{5}$ mile

$\dfrac{9}{10} = \dfrac{9}{10}$

$\dfrac{1}{2} = \dfrac{5}{10}$

$\dfrac{9}{10} - \dfrac{5}{10} = \dfrac{4}{10} = \dfrac{2}{5}$

7. $\dfrac{3}{8}$ of a pound

$\dfrac{1}{1} = \dfrac{8}{8}$

$\dfrac{5}{8} = \dfrac{5}{8}$

$\dfrac{8}{8} - \dfrac{5}{8} = \dfrac{3}{8}$

8. $\dfrac{3}{5}$ inch

$\dfrac{2}{3} = \dfrac{10}{15}$

$\dfrac{1}{15} = \dfrac{1}{15}$

$\dfrac{10}{15} - \dfrac{1}{15} = \dfrac{9}{15} = \dfrac{3}{5}$

9. $\frac{1}{4}$ of her 1-hour lunch

$$\frac{1}{1} = \frac{4}{4}$$

$$\frac{3}{4} = \frac{3}{4}$$

$$\frac{4}{4} - \frac{3}{4} = \frac{1}{4}$$

10. $\frac{3}{8}$ acre

$$\frac{3}{4} = \frac{6}{8}$$

$$\frac{3}{8} = \frac{3}{8}$$

$$\frac{6}{8} - \frac{3}{8} = \frac{3}{8}$$

LESSON 25 ADDING MIXED NUMBERS

Practice

1. 11

$$\frac{1}{2} + \frac{1}{2} = \frac{2}{2}$$

$$\frac{2}{2} = 1$$

$$1 + 9 + 1 = 11$$

$3\frac{19}{24}$

$$\frac{2}{3} = \frac{16}{24}$$

$$\frac{1}{8} = \frac{3}{24}$$

$$\frac{16}{24} + \frac{3}{24} = \frac{19}{24}$$

$$2 + 1 = 3$$

$$3 + \frac{19}{24} = 3\frac{19}{24}$$

$5\frac{19}{20}$

$$\frac{3}{4} = \frac{15}{20}$$

$$\frac{1}{5} = \frac{4}{20}$$

$$\frac{15}{20} + \frac{4}{20} = \frac{19}{20}$$

$$5 + \frac{19}{20} = 5\frac{19}{20}$$

2. $3\frac{5}{8}$

$$\frac{1}{2} = \frac{4}{8}$$

$$\frac{1}{8} = \frac{1}{8}$$

$$\frac{4}{8} + \frac{1}{8} = \frac{5}{8}$$

$$3 + \frac{5}{8} = 3\frac{5}{8}$$

$8\frac{7}{8}$

$$\frac{1}{8} = \frac{1}{8}$$

$$\frac{3}{4} = \frac{6}{8}$$

$$\frac{1}{8} + \frac{6}{8} = \frac{7}{8}$$

$$8 + \frac{7}{8} = 8\frac{7}{8}$$

5

$$\frac{1}{2} + \frac{1}{2} = \frac{2}{2}$$

$$\frac{2}{2} = 1$$

$$2 + 2 + 1 = 5$$

There are many possible ways to estimate the answers in 3.
Just be sure that your answers are close to the ones given here.

3. 8

$5\frac{1}{10}$ rounds to 5

$2\frac{4}{5}$ rounds to 3

7

$4\frac{7}{8}$ rounds to 5

$1\frac{9}{10}$ rounds to 2

4

$2\frac{1}{8}$ rounds to 2

$2\frac{1}{7}$ rounds to 2

4. $11\frac{5}{8}$ miles

$$10\frac{1}{2} = 10\frac{4}{8}$$

$$1\frac{1}{8} = 1\frac{1}{8}$$

$$10\frac{4}{8} + 1\frac{1}{8} = 11\frac{5}{8}$$

5. $13\frac{1}{2}$ feet

$$2\frac{1}{2} = 2\frac{2}{4}$$

$$4\frac{1}{4} = 4\frac{1}{4}$$

$$2\frac{2}{4} + 2\frac{2}{4} + 4\frac{1}{4} + 4\frac{1}{4} = 12\frac{6}{4}$$

$$12\frac{6}{4} = 13\frac{2}{4} = 13\frac{1}{2}$$

LESSON 26   SUBTRACTING MIXED NUMBERS

Practice

1. $4\frac{7}{8}$

$$7\frac{1}{8} = 7\frac{1}{8}$$

$$2\frac{1}{4} = 2\frac{2}{8}$$

$$7\frac{1}{8} = 6\frac{8}{8} + \frac{1}{8} = 6\frac{9}{8}$$

$$6\frac{9}{8} - 2\frac{2}{8} = 4\frac{7}{8}$$

$6\frac{3}{10}$

$$9\frac{9}{10} = 9\frac{9}{10}$$

$$3\frac{3}{5} = 3\frac{6}{10}$$

$$9\frac{9}{10} - 3\frac{6}{10} = 6\frac{3}{10}$$

$5\frac{8}{9}$

$$6\frac{1}{3} = 6\frac{3}{9}$$

$$\frac{4}{9} = \frac{4}{9}$$

$$6\frac{3}{9} = 5\frac{9}{9} + \frac{3}{9} = 5\frac{12}{9}$$

$$5\frac{12}{9} - \frac{4}{9} = 5\frac{8}{9}$$

2. $3\frac{3}{8}$

$$3\frac{1}{2} = 3\frac{4}{8}$$

$$\frac{1}{8} = \frac{1}{8}$$

$$3\frac{4}{8} - \frac{1}{8} = 3\frac{3}{8}$$

$7\frac{3}{8}$

$$8\frac{1}{8} = 8\frac{1}{8}$$

$$\frac{3}{4} = \frac{6}{8}$$

$$8\frac{1}{8} = 7\frac{8}{8} + \frac{1}{8} = 7\frac{9}{8}$$

$$7\frac{9}{8} - \frac{6}{8} = 7\frac{3}{8}$$

$$\frac{3}{4}$$

$$3\frac{1}{2} = 3\frac{2}{4}$$

$$2\frac{3}{4} = 2\frac{3}{4}$$

$$3\frac{2}{4} = 2\frac{4}{4} + \frac{2}{4} = 2\frac{6}{4}$$

$$2\frac{6}{4} - 2\frac{3}{4} = \frac{3}{4}$$

3. 1   8   15

$2\frac{9}{10}$ rounds to 3 $8\frac{3}{4}$ rounds to 9 $20\frac{1}{8}$ rounds to 20

$2\frac{1}{5}$ rounds to 2 $1\frac{1}{12}$ rounds to 1 $5\frac{1}{9}$ rounds to 5

4. $4\frac{5}{8}$ points   **5.** $70\frac{1}{5}$ miles

$$51\frac{3}{4} = 51\frac{6}{8} \qquad\qquad 445\frac{1}{10} = 444\frac{11}{10}$$

$$47\frac{1}{8} = 47\frac{1}{8} \qquad\qquad 444\frac{11}{10} - 374\frac{9}{10} = 70\frac{2}{10}$$

$$51\frac{6}{8} - 47\frac{1}{8} = 4\frac{5}{8} \qquad\qquad 70\frac{2}{10} = 70\frac{1}{5}$$

6. $6\frac{3}{4}$ gallons   **7.** $1\frac{3}{8}$ inches

$$10\frac{1}{2} = 10\frac{2}{4} = 9\frac{6}{4} \qquad\qquad 50\frac{1}{8} = 49\frac{9}{8}$$

$$3\frac{3}{4} = 3\frac{3}{4} \qquad\qquad 48\frac{3}{4} = 48\frac{6}{8}$$

$$9\frac{6}{4} - 3\frac{3}{4} = 6\frac{3}{4} \qquad\qquad 49\frac{9}{8} - 48\frac{6}{8} = 1\frac{3}{8}$$

LESSON 27   MULTIPLYING FRACTIONS

Practice

1. $\dfrac{1}{8}$ $\dfrac{3}{20}$

$$\frac{1}{2} \times \frac{1}{4} = \frac{1}{8} \qquad\qquad \frac{3}{\overset{}{\underset{4}{\cancel{8}}}} \times \frac{\overset{1}{\cancel{2}}}{5} = \frac{3}{4} \times \frac{1}{5} = \frac{3}{20}$$

$$\frac{1}{2}$$

$$\frac{\overset{1}{\cancel{5}}}{\underset{2}{\cancel{6}}} \times \frac{\overset{1}{\cancel{3}}}{\underset{1}{\cancel{5}}} = \frac{1}{2} \times \frac{1}{1} = \frac{1}{2}$$

2. $\dfrac{7}{12}$ $\qquad\qquad\qquad\qquad\qquad\qquad$ $\dfrac{3}{16}$

$\dfrac{\overset{1}{\cancel{2}}}{3} \times \dfrac{7}{\underset{4}{\cancel{8}}} = \dfrac{1}{3} \times \dfrac{7}{4} = \dfrac{7}{12}$ $\qquad\quad$ $\dfrac{3}{\underset{2}{\cancel{10}}} \times \dfrac{\overset{1}{\cancel{5}}}{8} = \dfrac{3}{2} \times \dfrac{1}{8} = \dfrac{3}{16}$

$\dfrac{1}{3}$

$\dfrac{\overset{1}{\cancel{4}}}{\underset{3}{\cancel{9}}} \times \dfrac{\overset{1}{\cancel{3}}}{\underset{1}{\cancel{4}}} = \dfrac{1}{3} \times \dfrac{1}{1} = \dfrac{1}{3}$

3. $1\dfrac{5}{8}$ $\qquad\qquad\qquad$ $\dfrac{27}{32}$ $\qquad\qquad\qquad$ $15\dfrac{1}{2}$

$3\dfrac{1}{4} = \dfrac{13}{4}$ $\qquad\qquad$ $1\dfrac{1}{8} = \dfrac{9}{8}$ $\qquad\qquad$ $5 = \dfrac{5}{1}$

$\dfrac{13}{4} \times \dfrac{1}{2} = \dfrac{13}{8}$ $\qquad\quad$ $\dfrac{9}{8} \times \dfrac{3}{4} = \dfrac{27}{32}$ $\qquad\quad$ $3\dfrac{1}{10} = \dfrac{31}{10}$

$\dfrac{13}{8} = 1\dfrac{5}{8}$ $\qquad\qquad\qquad\qquad\qquad\qquad$ $\dfrac{\overset{1}{\cancel{5}}}{1} \times \dfrac{31}{\underset{2}{\cancel{10}}} = \dfrac{1}{1} \times \dfrac{31}{2} = 15\dfrac{1}{2}$

4. $1\dfrac{1}{9}$ $\qquad\qquad\qquad\qquad\qquad\qquad$ $1\dfrac{11}{25}$

$\dfrac{5}{6} \times 1\dfrac{1}{3} = \dfrac{5}{6} \times \dfrac{4}{3}$ $\qquad\qquad$ $\dfrac{9}{10} \times 1\dfrac{3}{5} = \dfrac{9}{10} \times \dfrac{8}{5}$

$\dfrac{5}{\underset{3}{\cancel{6}}} \times \dfrac{\overset{2}{\cancel{4}}}{3} = \dfrac{5}{3} \times \dfrac{2}{3} = \dfrac{10}{9}$ $\qquad$ $\dfrac{9}{\underset{5}{\cancel{10}}} \times \dfrac{\overset{4}{\cancel{8}}}{5} = \dfrac{9}{5} \times \dfrac{4}{5} = \dfrac{36}{25}$

$\dfrac{10}{9} = 1\dfrac{1}{9}$ $\qquad\qquad\qquad\qquad$ $\dfrac{36}{25} = 1\dfrac{11}{25}$

$\dfrac{3}{4}$

$\dfrac{2}{3} \times 1\dfrac{1}{8} = \dfrac{2}{3} \times \dfrac{9}{8}$

$\dfrac{\overset{1}{\cancel{2}}}{\underset{1}{\cancel{3}}} \times \dfrac{\overset{3}{\cancel{9}}}{\underset{4}{\cancel{8}}} = \dfrac{1}{1} \times \dfrac{3}{4} = \dfrac{3}{4}$

5. $\frac{7}{8}$ cup

$$1\frac{3}{4} = \frac{7}{4}$$

$$\frac{7}{4} \times \frac{1}{2} = \frac{7}{8}$$

6. $64\frac{1}{2}$ miles

$$10\frac{3}{4} = \frac{43}{4}$$

$$6 = \frac{6}{1}$$

$$\frac{43}{\overset{}{\underset{2}{4}}} \times \frac{\overset{3}{6}}{1} = \frac{129}{2} = 64\frac{1}{2}$$

7. 54 feet

$$4\frac{1}{2} = \frac{9}{2}$$

$$12 = \frac{12}{1}$$

$$\frac{9}{\underset{1}{\overset{}{2}}} \times \frac{\overset{6}{12}}{1} = \frac{54}{1} = 54$$

8. $107\frac{1}{3}$ square yards

$$9\frac{1}{3} = \frac{28}{3}$$

$$11\frac{1}{2} = \frac{23}{2}$$

$$\frac{14}{3} \times \frac{23}{1} = \frac{322}{3} = 107\frac{1}{3}$$

LESSON 28   DIVIDING FRACTIONS

Practice

1. $1\frac{1}{2}$

$$\frac{3}{4} \div \frac{1}{2} = \frac{3}{4} \times \frac{2}{1}$$

$$\frac{3}{\underset{2}{\overset{}{4}}} \times \frac{\overset{1}{2}}{1} = \frac{3}{2} \times \frac{1}{1}$$

$$\frac{3}{2} \times \frac{1}{1} = \frac{3}{2}$$

$$\frac{3}{2} = 1\frac{1}{2}$$

$1\frac{1}{3}$

$$\frac{5}{6} \div \frac{5}{8} = \frac{5}{6} \times \frac{8}{5}$$

$$\frac{\overset{1}{5}}{\underset{3}{\overset{}{6}}} \times \frac{\overset{4}{8}}{\underset{1}{\overset{}{5}}} = \frac{1}{3} \times \frac{4}{1}$$

$$\frac{1}{3} \times \frac{4}{1} = \frac{4}{3}$$

$$\frac{4}{3} = 1\frac{1}{3}$$

$2\frac{1}{2}$

$$\frac{5}{8} \div \frac{1}{4} = \frac{5}{8} \times \frac{4}{1}$$

$$\frac{5}{\underset{2}{\overset{}{8}}} \times \frac{\overset{1}{4}}{1} = \frac{5}{2} \times \frac{1}{1}$$

$$\frac{5}{2} \times \frac{1}{1} = \frac{5}{2}$$

$$\frac{5}{2} = 2\frac{1}{2}$$

2. 8

$$4 \div \frac{1}{2} = 4 \times \frac{2}{1}$$

$$4 = \frac{4}{1}$$

$$\frac{4}{1} \times \frac{2}{1} = \frac{8}{1}$$

$$\frac{8}{1} = 8$$

40

$$5 \div \frac{1}{8} = 5 \times \frac{8}{1}$$

$$5 = \frac{5}{1}$$

$$\frac{5}{1} \times \frac{8}{1} = \frac{40}{1}$$

$$\frac{40}{1} = 40$$

$\frac{5}{24}$

$$\frac{5}{8} \div 3 = \frac{5}{8} \times \frac{1}{3}$$

$$\frac{5}{8} \times \frac{1}{3} = \frac{5}{24}$$

3. $1\frac{3}{8}$

$2\frac{3}{4} \div 2 = 2\frac{3}{4} \times \frac{1}{2}$

$2\frac{3}{4} = \frac{11}{4}$

$\frac{11}{4} \times \frac{1}{2} = \frac{11}{8}$

$\frac{11}{8} = 1\frac{3}{8}$

$5\frac{1}{2}$

$1\frac{5}{6} \div \frac{1}{3} = 1\frac{5}{6} \times \frac{3}{1}$

$1\frac{5}{6} = \frac{11}{6}$

$\frac{11}{\cancel{6}_{2}} \times \frac{\cancel{3}^{1}}{1} = \frac{11}{2} \times \frac{1}{1}$

$\frac{11}{2} \times \frac{1}{1} = \frac{11}{2}$

$\frac{11}{2} = 5\frac{1}{2}$

$6\frac{1}{2}$

$2\frac{3}{5} \div \frac{2}{5} = 2\frac{3}{5} \times \frac{5}{2}$

$2\frac{3}{5} = \frac{13}{5}$

$\frac{13}{\cancel{5}_{1}} \times \frac{\cancel{5}^{1}}{2} = \frac{13}{1} \times \frac{1}{2}$

$\frac{13}{1} \times \frac{1}{2} = \frac{13}{2}$

$\frac{13}{2} = 6\frac{1}{2}$

4. $7\frac{1}{2}$

$1\frac{7}{8} \div \frac{1}{4} = 1\frac{7}{8} \times \frac{4}{1}$

$1\frac{7}{8} = \frac{15}{8}$

$\frac{15}{\cancel{8}_{2}} \times \frac{\cancel{4}^{1}}{1} = \frac{15}{2} \times \frac{1}{1}$

$\frac{15}{2} \times \frac{1}{1} = \frac{15}{2}$

$\frac{15}{2} = 7\frac{1}{2}$

$1\frac{11}{14}$

$4\frac{1}{6} = \frac{25}{6}$

$2\frac{1}{3} = \frac{7}{3}$

$\frac{25}{6} \div \frac{7}{3} = \frac{25}{6} \times \frac{3}{7}$

$\frac{25}{\cancel{6}_{2}} \times \frac{\cancel{3}^{1}}{7} = \frac{25}{2} \times \frac{1}{7}$

$\frac{25}{2} \times \frac{1}{7} = \frac{25}{14}$

$\frac{25}{14} = 1\frac{11}{14}$

100

$$10 = \frac{10}{1}$$

$$\frac{10}{1} \div \frac{1}{10} = \frac{10}{1} \times \frac{10}{1}$$

$$\frac{10}{1} \times \frac{10}{1} = 100$$

5. Two segments

$$10\frac{1}{2} = \frac{21}{2}$$

$$5\frac{1}{4} = \frac{21}{4}$$

$$\frac{21}{2} \div \frac{21}{4} = \frac{\overset{1}{\cancel{21}}}{\underset{1}{\cancel{2}}} \times \frac{\overset{2}{\cancel{4}}}{\underset{1}{\cancel{21}}} = 2$$

6. $5\frac{7}{8}$ feet

$$23\frac{1}{2} = \frac{47}{2}$$

$$4 = \frac{4}{1}$$

$$\frac{47}{2} \div \frac{4}{1} = \frac{47}{2} \times \frac{1}{4} = \frac{47}{8}$$

$$\frac{47}{8} = 5\frac{7}{8}$$

7. $\frac{21}{50}$ meter

$$2\frac{1}{10} = \frac{21}{10}$$

$$5 = \frac{5}{1}$$

$$\frac{21}{10} \div \frac{5}{1} = \frac{21}{10} \times \frac{1}{5} = \frac{21}{50}$$

8. $21\frac{1}{3}$ steaks

$$16 = \frac{16}{1}$$

$$\frac{3}{4} = \frac{3}{4}$$

$$\frac{16}{1} \div \frac{3}{4} = \frac{16}{1} \times \frac{4}{3} = \frac{64}{3}$$

$$\frac{64}{3} = 21\frac{1}{3}$$

9. 6 service calls

$$9 = \frac{9}{1}$$

$$1\frac{1}{2} = \frac{3}{2}$$

$$\frac{9}{1} \div \frac{3}{2} = \frac{\overset{3}{\cancel{9}}}{1} \times \frac{2}{\underset{1}{\cancel{3}}} = \frac{6}{1}$$

$$\frac{6}{1} = 6$$

Unit 1 Review

1.

$$\begin{array}{r} 409 \\ +\ 201 \\ \hline 610 \end{array}$$

$$\begin{array}{r} 1,024 \\ -\ \ \ 14 \\ \hline 1,010 \end{array}$$

$$\begin{array}{r} 138 \\ +\ 328 \\ \hline 466 \end{array}$$

$$\begin{array}{r} 560 \\ -\ 245 \\ \hline 315 \end{array}$$

2.

$$\begin{array}{r} 907 \\ +\ \ 458 \\ \hline 1,365 \end{array}$$

$$\begin{array}{r} 265 \\ -\ \ 95 \\ \hline 170 \end{array}$$

$$\begin{array}{r} 2,209 \\ +\ 1,032 \\ \hline 3,241 \end{array}$$

3.

$$\begin{array}{r} 237 \\ -\ 125 \\ \hline 112 \end{array}$$

$$\begin{array}{r} 218 \\ +\ 136 \\ \hline 354 \end{array}$$

$$\begin{array}{r} 4,010 \\ -\ 2,400 \\ \hline 1,610 \end{array}$$

4.

$$\begin{array}{r} 45 \\ \times\ \ 18 \\ \hline 360 \\ 45\ \ \\ \hline 810 \end{array}$$

$$\begin{array}{r} 308 \\ \times\ \ 32 \\ \hline 616 \\ 9\ 24\ \ \\ \hline 9,856 \end{array}$$

$$\begin{array}{r} 144 \\ \times\ \ 12 \\ \hline 288 \\ 1\ 44\ \ \\ \hline 1,728 \end{array}$$

$$\begin{array}{r} 312 \\ \times\ \ \ 60 \\ \hline 000 \\ 1,872\ \ \\ \hline 18,720 \end{array}$$

5. 10r5

$$\begin{array}{r} 10 \\ 15\overline{)155} \\ \underline{15} \\ 05 \end{array}$$

12

$$\begin{array}{r} 12 \\ 12\overline{)144} \\ \underline{12} \\ 24 \\ \underline{24} \\ 0 \end{array}$$

8r24

$$\begin{array}{r} 8 \\ 32\overline{)280} \\ \underline{256} \\ 24 \end{array}$$

200

$$\begin{array}{r} 200 \\ 50\overline{)10,000} \\ \underline{10\ 0} \\ 00 \\ \underline{0} \\ 00 \\ \underline{0} \\ 0 \end{array}$$

6. 2,640

$$\begin{array}{r} 88 \\ \times\ \ 30 \\ \hline 00 \\ 2\ 64\ \ \\ \hline 2,640 \end{array}$$

11r10

$$\begin{array}{r} 11 \\ 15\overline{)175} \\ \underline{15} \\ 25 \\ \underline{15} \\ 10 \end{array}$$

225

$$\begin{array}{r} 25 \\ \times\ \ \ 9 \\ \hline 225 \end{array}$$

7. 4r25

$$\begin{array}{r} 4 \\ 30\overline{)145} \\ \underline{120} \\ 25 \end{array}$$

8,160

$$\begin{array}{r} 160 \\ \times\ \ 51 \\ \hline 160 \\ 8\ 00\ \ \\ \hline 8,160 \end{array}$$

16

$$\begin{array}{r} 16 \\ 16\overline{)256} \\ \underline{16} \\ 96 \\ \underline{96} \\ 0 \end{array}$$

There are many possible ways to estimate answers 8 to 10.
Just be sure that your answers are close to the ones given here.

8. 130
    48 rounds to 50
    79 rounds to 80

    190
    201 rounds to 200
    9 rounds to 10

    120
    97 rounds to 100
    21 rounds to 20

9. 400
    41 rounds to 40
    9 rounds to 10

    20
    198 rounds to 200
    11 rounds to 10

    3,000
    25 rounds to 30
    99 rounds to 100

10. 120
    15 rounds to 20
    109 rounds to 100

    4,000
    203 rounds to 200
    19 rounds to 20

    5
    97 rounds to 100
    19 rounds to 20

11. 21   68,913   424

12. 422   5,680   2,375

13. 320 minutes
    $16 \times 20 = 320$

14. 360 miles
    $148 + 212 = 360$

15. 625 plants
    $25 \times 25 = 625$

16. 29 pounds
    $8 + 8 + 13 = 29$

17. \$38,310
    $\$85,900 - \$47,590 = \$38,310$

18. 38 beakers
    $114 \div 3 = 38$

19. 97 cans
    $148 - 51 = 97$

20. 42,240 feet
    $5,280 \times 8 = 42,240$

Unit 2 Review

1.
$$5.07 + 2.80 = 7.87$$
$$\overset{9}{\cancel{10}}.{}^{1}5 - 9.6 = .9$$
$$115.24 + 100.09 = 215.33$$
$$20.9 - 10.0 = 10.9$$

2.
$$2.009 + 5.450 = 7.459$$
$$12.50 + 3.04 = 15.54$$

3.
$$2.600 + .008 = 2.608$$
$$\overset{1}{10}.590 + 5.075 = 15.665$$

4.
$$5.9 - .6 = 5.3$$
$$\overset{3}{\cancel{4}}.{}^{1}003 - 2.100 = 1.903$$

5.
$$\overset{6}{\cancel{7}}.{}^{1}0 - 3.8 = 3.2$$
$$200.\overset{8}{\cancel{9}}\overset{9}{\cancel{9}}{}^{1}0 - .175 = 200.725$$

6.

2.8	40.1	3.01	.09
× 1.5	× .2	× 5	× .32
1 4 0	8.02	15.05	18
2 8			27
4.2 0			.0288

7. 20.2   900   30

$$
\begin{array}{r} 2\,0.2 \\ 3.5\,)\overline{70.7\,0} \\ 70 \\ \hline 0\,7\,0 \\ 7\,0 \\ \hline 0 \end{array}
\qquad
\begin{array}{r} 9\,00 \\ .09\,)\overline{81.00} \\ 81 \\ \hline 0\,00 \end{array}
\qquad
\begin{array}{r} 30 \\ .003\,)\overline{.090} \\ 9 \\ \hline 00 \end{array}
$$

8.

$\overset{1}{9}.9$	5.0 8	$\overset{3}{1}\,\overset{3}{6}.6$
× 2	× 4.5	× .0 5
19.8	2,5 4 0	.8 3 0
	2,0 3 2	
	2 2.8 6 0	

9. 3,300   .02   7

$$
\begin{array}{r} 33\,00 \\ .03\,)\overline{99.00} \\ 9 \\ \hline 09 \\ 9 \\ \hline 0\,00 \end{array}
\qquad
\begin{array}{r} .02 \\ 4\,)\overline{.08} \\ 8 \\ \hline 0 \end{array}
\qquad
\begin{array}{r} 7 \\ .8\,)\overline{5.6} \\ 5\,6 \\ \hline 0 \end{array}
$$

There are many possible ways to estimate answers 10 and 11. Just be sure that your answers are close to the ones given here.

10. 106   1   83
      5.9 rounds to 6   2.01 rounds to 2   59.7 rounds to 60
      100.1 rounds to 100   .9 rounds to 1   23.2 rounds to 23

11. 70   3   32
      6.6 rounds to 7   2.99 rounds to 3   4.009 rounds to 4
      10.1 rounds to 10   .8 rounds to 1   8.2 rounds to 8

12. 146.59   8.25

13. 545.16   245.7647

14. $2.20
      $4.40 ÷ 2 = $2.20

15. 20.75 ounces
      12.25 + 8.5 = 20.75

16. $173.40
      25.5 × $6.80 = $173.40

17. 26.67 centimeters
      2.54 × 10.5 = 26.67

18. 23.8 inches
      36.5 − 12.7 = 23.8

19. 22.4 miles
      4.7 + 2.8 + 6.5 + 4.5 + 3.9 = 22.4

20. 230.55 square yards
      15.9 × 14.5 = 230.55

Unit 3 Review

1. $\dfrac{2}{3}$ $\qquad\qquad$ $1\dfrac{1}{2}$ $\qquad\qquad$ $2\dfrac{1}{8}$

2. $\dfrac{35}{40} = \dfrac{7}{8}$ $\qquad$ **3.** $\dfrac{2}{18} = \dfrac{1}{9}$ $\qquad$ **4.** $\dfrac{5}{6}$ $\qquad$ **5.** $\dfrac{40}{100} = \dfrac{2}{5}$

6. $\dfrac{4}{5} = \dfrac{16}{20}$ $\qquad$ $\dfrac{2}{3} = \dfrac{12}{18}$ $\qquad$ $\dfrac{9}{10} = \dfrac{81}{90}$

7. $1\dfrac{1}{12}$ $\qquad\qquad\qquad$ $3\dfrac{1}{4}$ $\qquad\qquad\qquad$ $6\dfrac{3}{4}$

$\dfrac{1}{3} = \dfrac{4}{12}$ $\qquad\qquad$ $\dfrac{7}{8} + \dfrac{3}{8} = \dfrac{10}{8}$ $\qquad$ $2\dfrac{1}{4} + 4\dfrac{1}{2} = 6\dfrac{3}{4}$

$\dfrac{3}{4} = \dfrac{9}{12}$ $\qquad\qquad$ $1 + 1 = 2$

$\dfrac{4}{12} + \dfrac{9}{12} = \dfrac{13}{12}$ $\qquad$ $2 + \dfrac{10}{8} = 2\dfrac{10}{8}$

$\dfrac{13}{12} = 1\dfrac{1}{12}$ $\qquad\qquad$ $2\dfrac{10}{8} = 3\dfrac{2}{8}$

$\qquad\qquad\qquad\qquad$ $3\dfrac{2}{8} = 3\dfrac{1}{4}$

8. $\dfrac{3}{8}$ $\qquad\qquad\qquad$ $\dfrac{2}{3}$ $\qquad\qquad\qquad$ $3\dfrac{3}{8}$

$\dfrac{5}{8} = \dfrac{5}{8}$ $\qquad\qquad$ $1\dfrac{1}{3} = \dfrac{4}{3}$ $\qquad\qquad$ $\dfrac{1}{8} = \dfrac{1}{8}$

$\dfrac{1}{4} = \dfrac{2}{8}$ $\qquad\qquad$ $\dfrac{2}{3} = \dfrac{2}{3}$ $\qquad\qquad$ $\dfrac{3}{4} = \dfrac{6}{8}$

$\dfrac{5}{8} - \dfrac{2}{8} = \dfrac{3}{8}$ $\qquad$ $\dfrac{4}{3} - \dfrac{2}{3} = \dfrac{2}{3}$ $\qquad$ $6\dfrac{1}{8} - 2\dfrac{6}{8} = 5\dfrac{9}{8} - 2\dfrac{6}{8}$

$\qquad\qquad\qquad\qquad\qquad\qquad\qquad\qquad\qquad$ $5\dfrac{9}{8} - 2\dfrac{6}{8} = 3\dfrac{3}{8}$

9. $\dfrac{5}{12}$ $\qquad\qquad\qquad$ $29\dfrac{3}{4}$ $\qquad\qquad\qquad$ 11

$\dfrac{1}{2} \times \dfrac{5}{6} = \dfrac{5}{12}$ $\qquad$ $8\dfrac{1}{2} = \dfrac{17}{2}$ $\qquad\qquad$ $4 = \dfrac{4}{1}$

$\qquad\qquad\qquad\qquad$ $3\dfrac{1}{2} = \dfrac{7}{2}$ $\qquad\qquad$ $2\dfrac{3}{4} = \dfrac{11}{4}$

$\qquad\qquad\qquad\qquad$ $\dfrac{17}{2} \times \dfrac{7}{2} = \dfrac{119}{4}$ $\qquad$ $\dfrac{4}{1} \times \dfrac{11}{4} = \dfrac{1}{1} \times \dfrac{11}{1}$

$\qquad\qquad\qquad\qquad$ $\dfrac{119}{4} = 29\dfrac{3}{4}$ $\qquad\qquad$ $\dfrac{1}{1} \times \dfrac{11}{1} = 11$

10. $\frac{3}{8}$

$\frac{3}{4} \div 2 = \frac{3}{4} \times \frac{1}{2}$

$\frac{3}{4} \times \frac{1}{2} = \frac{3}{8}$

$3\frac{2}{5}$

$8\frac{1}{2} = \frac{17}{2}$

$2\frac{1}{2} = \frac{5}{2}$

$\frac{17}{2} \div \frac{5}{2} = \frac{17}{2} \times \frac{2}{5}$

$\frac{17}{\overset{}{\underset{1}{2}}} \times \frac{\overset{1}{2}}{5} = \frac{17}{1} \times \frac{1}{5}$

$\frac{17}{1} \times \frac{1}{5} = \frac{17}{5}$

$\frac{17}{5} = 3\frac{2}{5}$

8

$42 = \frac{42}{1}$

$5\frac{1}{4} = \frac{21}{4}$

$\frac{42}{1} \div \frac{21}{4} = \frac{42}{1} \times \frac{4}{21}$

$\frac{\overset{2}{42}}{1} \times \frac{4}{\underset{1}{21}} = \frac{2}{1} \times \frac{4}{1}$

$\frac{2}{1} \times \frac{4}{1} = \frac{8}{1}$

$\frac{8}{1} = 8$

11. 9

$3\frac{3}{4}$ rounds to 4

$5\frac{1}{8}$ rounds to 5

5

$9\frac{1}{10}$ rounds to 9

$3\frac{6}{7}$ rounds to 4

6

$2\frac{1}{5}$ rounds to 2

$3\frac{9}{10}$ rounds to 4

12. $30\frac{3}{4}$ feet

$10\frac{1}{4} = 10\frac{2}{8}$

$5\frac{1}{8} = 5\frac{1}{8}$

$10\frac{2}{8} + 10\frac{2}{8} + 5\frac{1}{8} + 5\frac{1}{8} = 30\frac{6}{8}$

$30\frac{6}{8} = 30\frac{3}{4}$

13. $52\frac{17}{32}$

$10\frac{1}{4} = \frac{41}{4}$

$5\frac{1}{8} = \frac{41}{8}$

$\frac{41}{4} \times \frac{41}{8} = \frac{1,681}{32}$

$\frac{1,681}{32} = 52\frac{17}{32}$

14. 27 miles

$4\frac{1}{2} = \frac{9}{2}$

$6 = \frac{6}{1}$

$\frac{9}{\underset{1}{2}} \times \frac{\overset{3}{6}}{1} = \frac{27}{1}$

$\frac{27}{1} = 27$

15. 12 truckloads

$37\frac{1}{2} = \frac{75}{2}$

$3\frac{1}{8} = \frac{25}{8}$

$\frac{75}{2} \div \frac{25}{8} = \frac{3}{\underset{1}{2}} \times \frac{\overset{4}{8}}{1} = \frac{12}{1}$

$\frac{12}{1} = 12$